CAMBRIDGE LIBRARY COLLECTION

Books of enduring scholarly value

Physical Sciences

From ancient times, humans have tried to understand the workings of the world around them. The roots of modern physical science go back to the very earliest mechanical devices such as levers and rollers, the mixing of paints and dyes, and the importance of the heavenly bodies in early religious observance and navigation. The physical sciences as we know them today began to emerge as independent academic subjects during the early modern period, in the work of Newton and other 'natural philosophers', and numerous sub-disciplines developed during the centuries that followed. This part of the Cambridge Library Collection is devoted to landmark publications in this area which will be of interest to historians of science concerned with individual scientists, particular discoveries, and advances in scientific method, or with the establishment and development of scientific institutions around the world.

Recollections and Reflections

Manchester-born Sir Joseph John Thomson (1858–1940), discoverer of the electron, was one of the most important Cambridge physicists of the later nineteenth and first half of the twentieth centuries. Succeeding Lord Rayleigh as Cavendish Professor of Experimental Physics, he directed the research interests of the laboratory, and eight of his students, including Rutherford, went on to win Nobel Prizes, as Thomson himself did in 1906. He was knighted in 1908, received the Order of Merit in 1912, and became Master of Trinity College in 1918. He also served as President of the Royal Society from 1915 from 1920 and was a government advisor on scientific research during World War I. This autobiography, published in 1936, covers all aspects of his career – his student days in Manchester, arrival in Cambridge, and growing international reputation. It gives a fascinating picture of Cambridge life and science at a dynamic period of development.

Cambridge University Press has long been a pioneer in the reissuing of out-of-print titles from its own backlist, producing digital reprints of books that are still sought after by scholars and students but could not be reprinted economically using traditional technology. The Cambridge Library Collection extends this activity to a wider range of books which are still of importance to researchers and professionals, either for the source material they contain, or as landmarks in the history of their academic discipline.

Drawing from the world-renowned collections in the Cambridge University Library, and guided by the advice of experts in each subject area, Cambridge University Press is using state-of-the-art scanning machines in its own Printing House to capture the content of each book selected for inclusion. The files are processed to give a consistently clear, crisp image, and the books finished to the high quality standard for which the Press is recognised around the world. The latest print-on-demand technology ensures that the books will remain available indefinitely, and that orders for single or multiple copies can quickly be supplied.

The Cambridge Library Collection will bring back to life books of enduring scholarly value (including out-of-copyright works originally issued by other publishers) across a wide range of disciplines in the humanities and social sciences and in science and technology.

Recollections and Reflections

JOSEPH JOHN THOMSON

CAMBRIDGE
UNIVERSITY PRESS

CAMBRIDGE UNIVERSITY PRESS

Cambridge, New York, Melbourne, Madrid, Cape Town,
Singapore, São Paolo, Delhi, Tokyo, Mexico City

Published in the United States of America by Cambridge University Press, New York

www.cambridge.org
Information on this title: www.cambridge.org/9781108037921

© in this compilation Cambridge University Press 2011

This edition first published 1936
This digitally printed version 2011

ISBN 978-1-108-03792-1 Paperback

RECOLLECTIONS
AND REFLECTIONS

SIR J. J. THOMSON
Reproduced by permission of The Royal Society from the painting by
Fiddes Watt in their possession.

RECOLLECTIONS
AND
REFLECTIONS

by

SIR J. J. THOMSON

O.M., D.Sc., F.R.S., etc.

Master of Trinity College, Cambridge

LONDON

G. BELL AND SONS, LTD.

1936

PRINTED IN GREAT BRITAIN
BY R. & R. CLARK, LIMITED, EDINBURGH

Preface

WHEN I rashly consented to write this book I did not realise how difficult it would be and how long it would take. I have never kept a diary and my memory has always been very patchy, good for things which are amusing, bad for those which are instructive. I have, therefore, had to spend a good deal of time in verifying dates and I cannot hope to have escaped mistakes. Though some of the work has been troublesome, much of it has been delightful. I have felt at times as if I were living my life over again, old friendships which meant much to me seem to have sprung again into life and, as is the way with memory, one recollection kindles others until they stretch without a gap from long ago to yesterday. I have realised even more vividly than I did before how fortunate I have been throughout my life. I have had good parents, good teachers, good colleagues, good pupils, good friends, great opportunities, good luck and good health.

My father died soon after I went to the Owens College, and had it not been for the sacrifices made by my mother I could not have completed the course there or come to Cambridge. What I owe to my teachers, colleagues and pupils is told in this book. The events which really determined my career—going to Owens College and coming to Trinity with all that it has meant to me—were sheer accidents. I went to Owens because my father happened to meet a friend who had heard of Owens

College—at that time very few had—and I came to Trinity because my teacher happened to be a Trinity man.

I was lucky again in my opportunities. New discoveries were made which supplied new and more powerful instruments for my researches just at the time I wanted them. Again, I have been blessed with remarkably good health. I began badly—for some months after I was born I was not expected to live—but since then I have very seldom been ill. I cannot remember any day in the last sixty years when my work has been interrupted through bad health.

Part of the account of the Cavendish Laboratory given in this book is taken from a chapter I wrote in 1910 for *The History of the Cavendish Laboratory*, a book which has been long out of print. The chapter, "Physics in My Time" does not profess to give an account of all the discoveries made between 1870 and the beginning of the war, a period which is one of the most prolific in the history of Physics and would have required not a chapter but several volumes. I have confined myself to those which had special connection with my own work or that of my pupils, or of which I saw something while they were in progress.

I wish to thank my wife and daughter for their valuable assistance in reading the proofs, my daughter for preparing the Name Index, and my publishers for much valuable advice and for the care they have taken in the preparation of this volume.

J. J. THOMSON

TRINITY LODGE
  CAMBRIDGE
November 1936

Contents

List of Illustrations

CHAPTER I

Boyhood and Owens College

I WAS born in Cheetham, a suburb of Manchester, on December 18, 1856; both time and place were fortunate, for the period between then and now has been one of the most eventful in the history of the world. From the beginning to the end, and especially in the latter half, there has been a quick succession of one stupendous event after another. Monarchies have fallen, and have been replaced by Republics and Dictatorships. Free trade, which as a Manchester man I naturally regarded for long as essential to the prosperity of the country, has gone too. But it is not of such things I shall speak in these reminiscences; more in my province are the inventions and discoveries made in the period which have wrought great changes in our social life. When I was a boy there were no bicycles, no motor cars, no aeroplanes, no electric light, no telephones, no wireless, no gramophones, no electrical engineering, no X-ray photographs, no cinemas and no germs, at least none recognised by the doctors.

It was a mere accident that I became a physicist; it was intended that I should be an engineer. In those days the only way of entering this profession was to be apprenticed to some firm of engineers, paying a considerable premium for the privilege. It was arranged that I should be apprenticed to Sharp-Stewart & Co., who had a great reputation as makers of locomotives, but they told my father that they had a long waiting list, and it would be

I   B

some time before I could begin work. My father happened to mention this to a friend, who said, "If I were you, instead of leaving the boy at school I should send him while he is waiting to the Owens College : it must be a pretty good kind of place, for young John Hopkinson who has just come out Senior Wrangler at Cambridge was educated there ". My father took this advice, and I went to the College.

This accident, which I regard as the most critical event in my life and which determined my career, could not have happened in any English provincial town except Manchester, for there was no other which had anything corresponding to Owens College. And it could not happen now in any town in England, though now there are many which have such Colleges. I was only fourteen when I went to Owens College, and I believe that sixteen is now the minimum age at which students are admitted to such Colleges. Indeed the authorities at Owens College thought my admission such a scandal—I expect they feared that students would soon be coming in perambulators—that they passed regulations raising the minimum age for admission, so that such a catastrophe should not happen again.

The first thing I remember about my education was being sent when a very small boy to dancing classes. I remember them though it is more than seventy years ago, because I hated them so intensely ; after a short time, however, the attempt to teach me dancing was given up as hopeless. After going for a year or two to a small school for young boys and girls, kept by two maiden ladies who were friends of my mother, I went to a private day school kept by two brothers named Townsend, at Alms Hill, Cheetham, which was near to where we lived. It was a

2

school which was not affected by the new views on education which were just then coming in. We were taught Latin from the Eton Latin Grammar, where the rules for syntax and grammar were in Latin, and we learnt these by heart before we knew what they meant. This was by no means so absurd as it might seem, as they were very skilfully written for their purpose. Erasmus is said to have had a hand in them. They rhymed, which made them very easy to remember ; and they were very sonorous, which made it quite a pleasure to roll them off. When I went to the Owens College, where they used a more up-to-date Grammar—I think it was called the Public Schools Latin Primer—I thought it very pallid and arid, and much preferred the lusciousness of the other. When I removed my books to Trinity Lodge in 1918 I came across the copy I had used at school, which I had not opened for nearly fifty years, and found that when I once got started I could go on with the " *propria quae maribus* " and the " *as in presenti perfectum format in avi* ", without a slip. I think that nowadays not enough advantage is taken of the ease with which boys can learn things by heart, indeed some people seem to regard memory almost as something which ought to be apologised for. The way we were taught English was to make us learn by heart " purple patches " from Shakespeare, Byron, and Scott, and I think this a better way than just reading the notes at the end of school editions of the works of these authors, which is now not an unusual way of preparing for an examination. We did not write essays ; the only English " composition " we did was to write at the end of term a letter to our parents about what we had been doing. This was dictated by the master of the form. I remember one of them which began, " My dear Parents —England this day expects every man to do his duty

3

was once the signal prelude to a great naval victory"— and thinking it would never have occurred to me to begin in that way.

History was not much more than learning a large number of dates, geography a description of the boundaries, principal towns and rivers of various countries, with a little map-drawing. In mathematics we learnt the propositions in the first book of Euclid by heart ; we did not do any algebra, but a great deal of arithmetic which was well taught. I think it forms the best introduction to mathematical ideas and is an excellent intellectual gymnastic, for it is easy to set simple questions which cannot be solved by rule of thumb, but require thought. Our knowledge was prevented from getting rusty by an ingenious device : lessons stopped a quarter of an hour before school ended, then the master would ask the boy at the head of the form some question—it might be a date in history, a sum in mental arithmetic, the Latin for an English word or the English for a Latin one and so on ; if his answer was right he could leave at once ; if not, the question went on to the next boy, and so on until someone answered it correctly and went out. A new question was then put to the boy next to him, and this process was repeated. A boy could not get out before the end of school unless he answered a question correctly. We had to do preparation at home, which generally took about an hour and always included something which had to be written out, a piece of Latin translation, sums or a piece of poetry—great attention was paid to the handwriting in these, and we got very few marks if this was not satisfactory. There were no examinations, and the prizes were awarded on the work done during the year. On the whole, I think I did not learn less under this old-fashioned system than I should under a modern one, but then, as I have said, I left the school when I was very young.

As for games, the older boys played cricket and football and a number of minor games such as rounders, prisoners' base, and a crude form of hockey called shinty which had no rule—or, if there was one, no one paid any attention to it—against raising the club, and I remember getting hit in the eye by a full swing. Fortunately it did not injure my sight. Football then was rather a gruesome business ; " hacking " was allowed, that is, the forwards in the scrum just kicked at anything in front of them, whether it was the ball or the shins of the opponents ; the result was that after a game one's shins were always bruised and often bleeding. We also played single-stick, an excellent game both as a gymnastic and as a training in keeping one's temper under provocation, as a hit over the legs arouses emotions which it requires practice to suppress.

Towards the end of my schooldays, the war between France and Germany broke out ; at first our sympathies were all with Germany ; we thought, like many people wiser than ourselves, that poor Germany would have no chance against the great army of France. It was not very long, however, before our sympathies were all the other way.

When I was a boy I was, as I am now, very fond of gardening, and had a little garden in which I was allowed to do as I liked. I spent the greater part of my pocket-money on a weekly gardening paper—there were no penny ones in those days. This paper was the ruin of my garden, for from time to time it gave such alluring accounts of some plant I had not got, that I felt I must buy a packet of seed and try to grow it. My garden was so small that the only way I could find room for this was to pull up something already growing, so that most of my plants came to an untimely end. I was also, as I have been all my life, fond of searching for wild flowers and of reading books about them, and thought that when I grew up I should like to be a botanist. The scientific names of plants irritated me greatly ; it was the days when the Linnæan system was almost universal, and I determined that if I became a botanist I would do all I could to purge botany from science.

Natural history was pressed into the service of boys when I and some of my friends tried to raise money to buy wickets, bats and balls for a small cricket club we wanted to start, by having an exhibition of butterflies, moths, birds' eggs and wild flowers which our relatives were expected to attend, and pay for admission. This worked quite well financially, but there was an unfortunate incident. An aunt of mine, looking at an exhibit of a big moth with a pin stuck through it, saw that the moth was alive and trying to get away. She was very indignant and rated the boy soundly for not killing it before pinning it down ; he excused himself by saying, " Well if it's not dead, it's its own fault. I've been pinching its head all morning." Another occasion, when I got some fun as well as instruction from my excursions into science, was once when a friend came in as I was using a microscope my father had given me. I showed it to him, plucked a hair from my head and put it on the slide and told him to look at it ; he did and seemed very much interested, much more so than I had expected, for he was not very intelligent. He kept screwing it up and down. I thought perhaps the hair had been blown away. So I said, " Can you see it ? " " Oh yes," he said, " I can see it." " Doesn't it look very big ? " " It looks big enough, but I can't see the number on it." "Number," I said, "what number?" "Well," he said, "it says in the Bible that the hairs of our head are all numbered, but I can't find any number on this."

Manchester has played a prominent part in the history of physical science, for in it, in the first half of the nineteenth century, Dalton made the experiments which led him to the discovery of the law of multiple proportion in chemical combination, and Joule those which were instrumental in establishing the principle of the Conservation of Energy. Notable discoveries, though not of the same rank as these, were made by Sturgeon, the inventor of the electromagnet, and the Henrys, father and son, who did good work in physical chemistry.

The Manchester Literary and Philosophical Society is, I believe, the oldest, as it is certainly quite one of the most important, of such societies. In 1781 it began, as such societies generally did, by informal meetings of a few friends in each other's houses.[1] This developed into weekly meetings at a tavern ; then a house was built, and in 1800 John Dalton, the great chemist, was appointed Secretary. He had come to Manchester shortly before this as teacher of mathematics in the newly established Manchester College, an unsectarian College, which was the germ from which Manchester New College developed. His laboratory was in the house of the Society, and his diary is still in their possession. It was published in Roscoe and Harden's

---

[1] The only one of my connections who to my knowledge took any interest in scientific matters was G. V. Vernon, a cousin of my mother's, who was for some years one of the Honorary Secretaries of the Manchester Literary and Philosophical Society. He was a cotton spinner by trade— partner in the firm Vernon & Bazley, but a meteorologist by inclination, and published several papers on that subject.

New View of the Origin of Dalton's Atomic Theory, 1896.
It shows that Dalton, influenced by the views of Newton,
believed in the existence of atoms long before 1803, when
his great paper was read before the Philosophical Society,
and also that, contrary to the view previously held, Dalton
was not led to the discovery of the atomic theory by discovering the law of Chemical Combination in multiple
proportion, but was led by the atomic theory to the discovery of this law. Though not the author of the atomic
theory, he has strong claims to have been the first to prove
it, or at any rate to find evidence in its favour which, if
not absolutely conclusive, was immensely stronger than
any that had been obtained before. In 1817 he was
elected President of the Society, and remained so until his
death in 1844. He was not a fluent speaker, and when,
as President, he had to make a few remarks when the
reader of a paper stopped, he is reported to have sometimes
contented himself by saying, " This paper will no doubt
be found interesting by those who take an interest in it ".
  Though he was one of the shyest and most retiring of
men, the Manchester people knew that they had a great
man living among them, and were proud of it. They
showed their appreciation by raising in his lifetime a sum
of £2000 for a statue by Chantrey. On the day of his
funeral many of the mills and shops were closed ; and an
important street in the city is called John Dalton Street.
  Dalton was succeeded as President by James Prescott
Joule, who had been his pupil : he had refused, however, to
let him go on to chemistry until he had read mathematics.
Joule was only twenty-six when he became President,
but four years before that he had discovered the very
important law which bears his name, viz. that the heat
produced in a given time in a circuit through which a

current is passing is proportional to the product of the electrical resistance of the circuit and the square of the current. In 1843 he published the paper which contains the first measurement of the mechanical equivalent of heat, which, with his subsequent experiments on the same subject, were in the main responsible for the acceptance of the principle of the Conservation of Energy. This was a remarkable achievement for a young man only twenty-five years of age, and was a striking instance of how great generalisations may be reached by patient work. He began by supposing that heat was a substance and so could neither be created nor destroyed. He soon convinced himself that this could not be true ; he then measured the proportion between the heat produced when paddles were rotated in a vessel full of water, and the work required to rotate the paddle. He found that this did not depend on the kind of fluid in the vessel or the kind of machine used to churn it, and he came to the conclusion that whenever a certain amount of heat was produced a proportional amount of mechanical work must be spent. This did not meet with immediate acceptance. Even William Thomson, who subsequently became one of his closest friends and greatest admirers, at first believed that it must be wrong. Thomson was profoundly impressed by the power of the method introduced by Carnot in 1825, and called Carnot's cycle, in predicting new physical phenomena ; for example, it follows from it that, since water expands on freezing, the melting point of ice must be lowered by pressure ; or again, since the surface tension of a soap film diminishes when its temperature rises, the temperature of a film must fall when it is stretched. Carnot's cycle, in the form in which it was first published, involved the assumption that

9

heat could neither be created nor destroyed. Thomson, however, showed later that this assumption was not essential ; and Clausius had, a short time before, come to the same conclusion.

Joule, besides being an excellent experimenter, had a very clear mind and sound judgment. He was one of those physicists who have, I think, been more plentiful in this country than in any other, who, though not holding any Professorship or other official post, have devoted themselves to the advancement of science at their own cost and in their own laboratories. None of these have done work more important than Joule. When I was a boy I was introduced by my father to Joule, and when he had gone my father said, " Some day you will be proud to be able to say you have met that gentleman " ; and I am.

A statue of him by Gilbert stands side by side with that of Dalton in the Town Hall of Manchester.

John Owens, a Manchester merchant who died in 1846, left his estate to be devoted to the establishment of a College free from religious tests, for instruction in the branches of knowledge taught in the English universities. It was not, however, until 1851 that the College started ; the time was a bad one for realising any kind of property, and in addition there was a dispute as to what amount, if any, of religious instruction was consistent with the terms of the trust. Finally it was compromised by limiting the instruction to lectures on the Greek of the New Testament and the Hebrew of the Old. However, by March 1851 a Principal and a few Professors had been appointed and the College opened. The first year 62 students attended, 71 in the second, and then it began to decline, and fell to little more than half this number by 1857. The *Manchester Guardian*, in a leader about this time, said, "Explain it how we may, the fact is that the College is a mortifying failure ". The Principal, Dr. A. J. Scott, resigned in 1857 and the Professor of Classics, J. G. Greenwood, was appointed to succeed him, and remained Principal until 1889.

Dr. Scott, who had suffered from bad health during his tenure of office, was in many ways a very remarkable man. He had been Professor of English Language and Literature in University College, London, before coming to Manchester. Before that he had assisted Irving—the founder of the Catholic Apostolic Church—in his church in Regent Square, and a little later was deprived, for heresy, of his licence to preach in a Presbyterian church. He was the friend of Archdeacon Hare, Frederick Denison Maurice and George MacDonald. Not unnaturally his sympathies and interests were in the literary rather than the scientific activities of the College. The Manchester Grammar School satisfied to a very considerable extent the needs of literary students, and had the advantage of having a considerable number of closed scholarships at Brasenose College, Oxford, so that students who hoped to go to Oxford went to the School rather than to Owens College.

Under Principal Greenwood the prosperity of the College began slowly to improve, and that improvement has gone on without interruption from then to now. In its development to a university its name has twice been changed, first to a College of the newly formed and now defunct Victoria University, and then to Manchester University.

Owens College began in a house in Quay Street, Deansgate, which had once been the residence of Cobden. In his day Quay Street had been a fashionable residential quarter, but when I went to the College, in 1871, rather disreputable slums reached on one side almost up to its doors. The house was not a large one, and as by that time the number of students had increased to about 500, we were very much cramped for space. The Engineering Department was housed in what had been the stables. The stable itself was converted into the Lecture Room, and the hayloft above it into the Drawing Office, which had to be reached by an outside uncovered wooden staircase. The Chemical Department was more fortunate, as an adjacent house was used for the Laboratory. The only thing that could be called a Physical Laboratory was a room in which the apparatus used for Lecture experiments was stored. The cramped space was not without its advantage. We were so closely packed that it was very easy for us to get to know each other. Arts and Science students jostled against each other continually ; a crowd of mathematicians would be waiting outside a lecture room for it to discharge a Latin or Greek class. Thus one of the chief defects of non-residential colleges, the lack of opportunities for social intercourse between the students, was almost absent. Though I was only two years in Quay Street and three in the comparatively palatial buildings in Oxford Road, to which the College moved in 1872, my most vivid recollections are those of the old Quay Street days ; in Oxford Road we had more room but less company, and fewer opportunities of making friends.

Though Owens College was badly housed, no university in the country had a more brilliant staff of Professors. For Mathematics there was Thomas Barker, a Senior Wrangler and Fellow of Trinity College, Cambridge ; for Physics, Balfour Stewart ; for Engineering, Osborne Reynolds ; for Chemistry, H. E. Roscoe ; W. C. Williamson, the distinguished palaeo-botanist, took all Natural History for his province, and lectured on Botany, Zoology, and

Geology. There was Stanley Jevons for Logic and Political Economy. Adolphus Ward, who afterwards became Principal of Owens College, then Vice-Chancellor of Victoria University, then Master of Peterhouse and Vice-Chancellor of Cambridge University, was Professor of History and English Language ; and Bryce, who later became Lord Bryce and Ambassador at Washington, of Law.

Although he never published any papers on the subject, I have never known a better teacher of mathematics than Barker, and in some respects no one so good. His lectures were always very carefully prepared ; they were written out on the backs of old examination papers, and it was the belief of his pupils that he destroyed them as soon as they had been delivered, and prepared new ones each year. His object was to give us sound views on the fundamental principles of the various branches of mathematics rather than to train us in solving problems with great rapidity. It was not that we did no examples, for some were set at the end of each lecture, and in addition to the lectures there were two classes a week at which we did nothing but examples ; but the number of examples we did was small compared with that usually done in schools. Some of the subjects on which he lectured were not, I believe, taught then at any other university : for example, we had lectures on what might be called the logic of mathematics. The first Professor of Mathematics at Owens College, Archibald Sandeman, was a pioneer in this subject and had written a book on it called *Pelicotetics*. The text of this was as difficult as its name ; there were pages without a stop of any kind, and though some of us began it I never knew anyone who got far.

Another novelty in Barker's teaching was the introduction of Quaternions, a system of geometrical analysis introduced by Sir W. R. Hamilton. I should think I must be the only man living who learnt Quaternions before he had done any Analytical Geometry. Quaternions attracted a good deal of attention in the early seventies, largely through the enthusiastic support of Professor Tait of Edinburgh, who wrote an elementary textbook on the subject. Though the ideas introduced by Hamilton were very interesting and attractive, and though many physical laws, notably those of electrodynamics, are most concisely expressed in Quaternionic Notation, I always found that to solve a new problem in mathematical physics, unless it was of the simplest character, the older methods were more manageable and efficient. I do not suppose that the introduction of unusual subjects such as these in comparatively elementary courses is always to be recommended, but I think it was successful in this case and that we worked all the better for it ; we regarded ourselves as pioneers and that it was up to us to make good.

The thing nearest to Barker's heart, however, was not mathematics but mosses, on which he was an authority. He lived a very simple life, was not married, and had been fortunate in his investments, so that he was able to retire in middle life to Derbyshire and devote himself to mosses. He left his estate to found a chair in Cryptogamic Botany in the University of Manchester, into which Owens College had developed.

I received great benefit from the teaching of Mr A. T. Bentley, also a member of Trinity College, who was lecturer in mathematics. He took the Lower Junior Class, in which I was put when I entered the College. I found his teaching very attractive and it was he who first aroused my interest in mathematics. Besides being a good teacher, he had a vigorous and breezy way with him which kept us alert and made his lectures very popular.

As I was taking the engineering course, the Professor I had most to do with in my first three years at Owens was Osborne Reynolds, the Professor of Engineering. He was one of the most original and independent of men, and never did anything or expressed himself like anybody else. The result was that it was very difficult to take notes at his lectures, so that we had to trust mainly to Rankine's text-books. Occasionally in the higher classes he would forget all about having to lecture and, after waiting for ten minutes or so, we sent the janitor to tell him that the class was waiting. He would come rushing into the room pulling on his gown as he came through the door, take a volume of Rankine from the table, open it apparently at random, see some formula or other and say it was wrong. He then went up to the blackboard to prove this. He wrote on the board with his back to us, talking to himself, and every now and then rubbed it all out and said that was wrong. He would then start afresh on a new line, and so on. Generally, towards the end of the lecture, he would finish one which he did not rub out, and say that this proved that Rankine was right after all. This, though it did not increase our knowledge of facts, was interesting, for it showed the working of a very acute mind grappling with a new problem. This was very characteristic of his research work. He would often begin with an idea which, after he had worked at it for some time, turned out to be wrong ; he would then start off on some other idea which had occurred to him while working at the previous one, and if this turned out wrong he would start another, and so on until he found one which satisfied him, and this was pretty sure to be right. He often started off in the wrong direction but he got to the goal in the end. He had a way of his own of doing most things. When he took up a problem, he did not begin by making a bibliography and reading the literature about the subject, but thought it out for himself from the beginning before reading what others had written about it.

There is, I think, a good deal to be said for his method. Many people's minds are more alert when they are thinking than when they are reading, and less liable to accept a plausible hypothesis which will not bear criticism. The novelty in his method of approach made his papers very hard reading—in fact I think it is probable that some of them have never been read through by anyone. He could, however, be clear enough when addressing a popular audience. Some of his Friday evening discourses given at the Royal Institution are models of clear exposition expressed in terse and nervous English. His best-known research is the one on the flow of water through cylindrical tubes. When the flow is slow the motion of the water is quite steady and there are no eddies, but this changes when the rate of flow exceeds a certain limit, the flow then becomes " turbulent ", and the stream is full of eddies. Reynolds investigated the conditions which determine this change, and the constant which determines the rate of flow at which it occurs is known as Reynolds' constant. He made important discoveries in lubrication ; he was attracted by what may be called out-of-door physics, and wrote papers on the calming of waves by rain, on the singing of a kettle, and why sound travelling against the wind is not heard so well as when it travels with it. He made some very beautiful experiments on the behaviour of vortex rings in water. We owe to him the generally accepted theory of the Radiometer ; the most complete account of this is in his paper " On certain Dimensional Properties of Matter in the Gaseous State ", *Phil. Trans.*, 1879, Part II. This paper is very difficult reading, so much so that a severe criticism of certain parts of it by Professor G. F. Fitzgerald, *Phil. Mag.* [5] 11, 1881, p. 103, was shown by Reynolds to be based on a wrong interpretation of his meaning. He worked out with great success the consequences which follow from the fact that collections of equal spheres can be " piled " in different ways : that for example they may be arranged either as in Fig. A or Fig. B, and showed that a sack full of shot arranged as in Fig. B would be quite rigid, while if the shot were as in Fig. A it would be quite floppy.

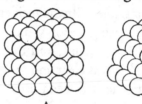

A   B

He called the subject dealing with effects of this character "dilatancy ". He was so impressed with the importance of it in connection with the structure of the universe that in his last publication, *The Sub-mechanics of the Universe*, he worked out a theory of the universe on the assumption that it was a collection of spheres in contact, and claimed that it was " the one, and the only one conceivable, purely mechanical system capable of accounting for all the physical evidence as we know it ". This is the most obscure of his writings, as at this time his mind was beginning to fail ; it is the record, however, of a great amount of work by a man of great originality and it is probable that, as Sir Horace Lamb said in his obituary notice of Osborne Reynolds, " a diligent study of it might bring to

light valuable results ". My personal relations with him
when I was a student are a very pleasant recollection ; he
was always very kind to me, had a winning way with him
and a charming smile.

My first introduction to physics was when I attended
in my second year at Owens the lectures on Elementary
Physics given by Professor Balfour Stewart ; these I found
very attractive and so clear that, young as I was, I had no
difficulty in understanding them. He looked a very old
man ; his hair was quite white, and I was very much surprised when I was told he was only forty-three. Quite
at the beginning of his work at Owens he had been badly
injured in a terrible railway accident, one of the worst on
record, at Abergele, North Wales. After about a year he
was able to resume work, and though he looked so much
older his mind was as clear and vigorous as before. When
a young man, he had been assistant to Professor David
Forbes at Edinburgh, and it was while at Edinburgh that
he made his most important contribution to physics—the
relation between the radiation and the absorption of radiation. In a paper published in the Edinburgh *Transactions*
vol. xxii. p. 1, March 1858, he enunciated the important
principle that in any enclosure bounded by opaque walls,
the radiation from any body of any kind of light must,
when the temperature throughout the enclosure is uniform,
be equal to the absorption of the same kind of light by
the body. Thus, if the body gives out light of a particular
wave-length it will absorb light of that wave-length, or if
a body like a plate of tourmaline absorbs light polarised in
one plane, it will, when radiating, give out light polarised
in one plane. The same principle was published about
a year and a half later by Kirchhoff, to whom it is gener-

ally attributed, as Stewart's work attracted at first little attention.[1]

Stewart was appointed as Director of the Kew Observatory in 1859, and worked mainly at Terrestrial Magnetism and the Periodicity of Sun Spots. Some friction on account of this arose between him and the Committee of the Royal Society which controlled the work of the Observatory, who thought he did not pay enough attention to routine observations. In consequence of this he welcomed the opportunity afforded by a vacancy in the Professorship of Physics at Owens, to escape into more congenial surroundings. He had in 1866 published an exceptionally able textbook on heat, which contains a very clear account of the laws of radiation.

In Quay Street the only approach to a Physical Laboratory was the room where the apparatus used for experiments was kept. When, however, the College removed in 1872 to the site which it still occupies under a new name, this want was supplied. The new laboratory, though it would seem almost microscopic in comparison with modern laboratories, was, I believe, the largest outside London at the time, and the first in which there were classes for practical work. These were not very largely attended, and the work of each student was not so rigidly prescribed as it has to be in the crowded classes of the present day. We were allowed considerable latitude in the choice of experiments. We set up the apparatus for ourselves and spent as much time as we pleased in investigating any point of interest that turned up in the course of our work. This was much more interesting and more educational than the highly organised systems which are necessary when the classes are large. Balfour Stewart was enthusiastic

[1] On this point see Lord Rayleigh, *Collected Papers*, vol. iv. p. 494.

about research, and succeeded in imparting the same spirit to some of his pupils. I remember, shortly after I began to work in the laboratory, he was talking to me about sun spots, and said he had made a large number of observations which he thought might throw some light on the connection between them and terrestrial phenomena, but that he had no time to reduce them. I ventured to say that if I could be of any help I should be glad to do what I could, and he gave me a number of observations to reduce. Though the work I did was purely arithmetical, I liked doing it and enjoyed the feeling that I was taking some part in real science. Some time after this, Stewart was trying to find out whether there is any change in weight when substances combine chemically. His method was to have some iodine at the bottom of a flask and to put in the flask a test-tube containing mercury ; the flask was then sealed up and its weight measured. The flask was then tilted, when the mercury ran out of the test-tube over the iodine and combined with it, and then the flask was weighed again. He asked me to make the weighings. I had made a good many without finding any difference or without there being an explosion of any kind. One Saturday afternoon, however, when I was alone in the laboratory, after tilting the flask, though the mercury ran over the iodine no combination took place. I held it up before my face to see what was the matter, when it suddenly exploded ; the hot compound of mercury and iodine went over my face and pieces of glass flew into my eyes. I managed to get out of the laboratory and found a porter, who summoned a doctor. For some days it was doubtful whether I should recover my sight. Mercifully I did so, and was able after a few weeks to get to work again. Modern physics teaches that a change of weight is produced by chemical combination, and enables us to predict what it should be. In the experiment I was trying, the change of weight would not be as great as one part in many thousand millions, so that it is not surprising that I did not detect it.

Before I left Manchester I did a small piece of research which was published in the *Proceedings* of the Royal Society.

Stewart in his lectures paid special attention to the principle of the Conservation of Energy, and gave a course of lectures entirely on this subject, and naturally I puzzled my head a great deal about it, especially about the transformation of one kind of energy into another —kinetic energy into potential energy, for example. I found the idea of kinetic energy being transformed into something of quite a different nature very perplexing, and it seemed to me simpler to suppose that all energy was of the same kind, and that the " transformation " of energy could be more correctly described as the transference of kinetic energy from one home to another, the effects it produced depending on the nature of its home. This had been recognised in the case of the transformation of the kinetic energy of a moving body striking against a target into heat, the energy of the heated body being the kinetic energy of its molecules, and it seemed to me that the same thing might apply to other kinds of energy. One day I plucked up courage to bring this view before Stewart. I should not have been surprised if he had regarded me as a heretic of the worst kind, and upbraided me for having profited so little from his teaching. He was, however, quite sympathetic. He did not profess to agree with it, but thought it was not altogether irrational, and that it might be worth my while to develop it.

Stewart had a strong turn for metaphysics, and, in conjunction with P. G. Tait, published in 1875 *The Unseen Universe*, which was an attempt to find a physical basis for immortality. This attracted a great deal of attention, and a second edition was called for in a few weeks. The physical basis was " Thought conceived to affect the matter of another universe simultaneously with this may explain a future State ". The authors attached so much importance to this that, to secure priority, they had taken the unusual step of publishing it as an anagram in *Nature* some months before the publication of their book. *The Unseen Universe* was followed by *Paradoxical Philosophy*, but this was not nearly so successful.

Towards the end of my stay at Owens, Arthur Schuster was also working in the laboratory and gave a course of lectures which I attended on Maxwell's *Treatise on Electricity and Magnetism*, which had just been published.

It was at this time that J. H. Poynting returned to Manchester after taking his degree as third Wrangler at Cambridge, and the friendship, which lasted until his death in 1914 and which was one of the greatest joys of my life, began. He had exceptionally sound judgment, a very original and acute mind, and a discussion with him always clarified my old ideas, often suggested new ones. In his paper on the transmission of energy in an electromagnetic field he introduced entirely novel ideas about the path the energy takes. The vector which represents this path, one of the most important in electrodynamic theory, is called the Poynting vector.

To his friends he was much more than a physicist, for he had a genius for friendship and a sympathy so delicate and acute that, whether you were well or ill, in high spirits or low, his company was a comfort and a delight. During a friendship of forty years I never saw him angry or impatient, and never heard him say an unkind thing about man, woman or child.

No one did more for the development of Owens College than Professor H. E. Roscoe, who succeeded Frankland as Professor of Chemistry in 1857, and held the Professorship until his retirement in 1885, when he was elected Member of Parliament for the Southern Division of Manchester. When he became Professor, Owens College was poverty-stricken, badly housed, with very few students. When he left, it was a university with very considerable funds, fine buildings, and more students than any other similar institution outside London. This was in the main due to Roscoe and Adolphus Ward, who was then the Professor of History, and subsequently Principal of the College. With the aid of Thomas Ashton, a rich cotton spinner who took a great interest in the College, they aroused the interest and obtained the assistance of many of the wealthy citizens of Manchester, who supplied sufficient funds to make these developments possible.

Roscoe was pre-eminently a man of affairs, with a very attractive and affable personality, sound judgment, energetic, persuasive, and very fertile in devising schemes. As an example of this, when during the early years of his Professorship, the Civil War in America had, by stopping the supply of cotton, thrown thousands of workpeople out of employment through no fault of their own, he with two others organised for their benefit a series of evening entertainments of various kinds. Some of these were concerts, some lectures on science with many experiments.

This scheme was a great success, the audiences averaging 4000 a week during the four winter months in which they were given. Encouraged by the success of these lectures, he instituted in 1866 a series of penny lectures ; a penny was charged for admission, and each lecture was published and sold at that price. He began by giving all the lectures himself, but he was soon able to enlist the services of Huxley, Tyndall, Sir John Lubbock, Balfour Stewart and many others. At first the lectures were all given in Manchester, but later he lectured at other manufacturing towns in Lancashire and Yorkshire. He took with him his lecture assistant, Heywood, and the apparatus required for the lecture experiments was in a large case. The railway porters at the stations around Manchester believed that he was running a Punch and Judy show. After a time the scheme of lectures was put on a more permanent basis by the Gilchrist Trustees ; later still the universities took a part, and provided short courses of study in various subjects, literary as well as scientific, in many towns. Roscoe, however, had for the first eleven years undertaken all the arrangements connected with these lectures.

They were the beginnings of a forty years' campaign to make the public realise the importance of science, to get more science taught in schools, to make science play a larger part in our industries, and to persuade the Government to make grants to universities and colleges for teaching and research. He was tireless in his efforts ; he wrote articles, he made speeches, he served on many Royal Commissions and took a very active part in the preparation of their reports, which attracted great attention and produced a very marked effect on public opinion.

He was elected M.P. for South Manchester in 1885. His selection as the candidate was very striking evidence of his popularity. The party managers had selected another candidate whom Roscoe was to propose, but at the public meeting to choose a candidate someone at the beginning of the proceedings suggested his name. This was received with great enthusiasm and carried unanimously. He was much the most effective representative science has ever had. His opinion carried great weight in the House : he had no axe to grind, he was not a manufacturer and he had given up his Professorship on entering Parliament, but he had the confidence both of manufacturers and of teachers. He was popular and worked hard, with the result that the scientific aspects of the Bills before Parliament received adequate consideration. He lost his seat at the election in 1895, when he was beaten by 76 votes by the Marquis of Lorne, and did not make any attempt to enter Parliament again. He had done a great work : it is hardly an exaggeration to say that the recognition of the importance of science for the welfare of the nation hardly existed when he began it ; teaching of science in schools was described in a report of a Royal Commission as being " regarded with jealousy by the staff, with contempt by the boys, and with indifference by the parents ". Now some science is taught in all schools and a good deal in a great many. Most of them have physical and chemical laboratories, which are in many cases large, well designed and well equipped with apparatus. There were then none of the Schools of Technology which are now to be found in most manufacturing towns ; some of these are magnificent buildings, erected by the municipalities at great cost, and provided with the most expensive apparatus. There were no Government institutions like the National Physical Laboratory for research, both in pure physics and for solving difficulties which manufacturers meet with in the course of their business. There were no laboratories for research in problems of importance to the army, navy or air force, such as are now to be found at Woolwich, Teddington and Farnborough, nor any of those started and subsidised by the Board of Scientific and Industrial Research, like the Fuel Board, the Food Board, and Research Laboratories in most of the industries of the country.

There were no scholarships for enabling students to take a course of research after taking a degree at a university. Then no one, except he held a teaching post or had private means, could undertake research, as there was no money to be made by it. It is very different now, when research is a recognised profession and a fairly lucrative one.

Besides the laboratories subsidised and originated by Government departments, many great firms, acting on the principle Heaven helps those who help themselves, have instituted research departments of their own, with large laboratories with men of great scientific ability at their head, and with a large staff of workers who have been trained at the universities in research. Imperial Chemical Industries and the General Electric Company are conspicuous examples of this. Another instance, in which a pupil of mine took part, is that of Messrs Tootal, Broadhurst Lee Co., Manchester ; they are makers of cotton goods and they realised how greatly the use of these would be extended if, like woollen goods, they did not crease.

If you squeeze a cotton handkerchief up in your hand, the creases remain after your hand is taken away, while if you squeeze a woollen cloth the creases will come out automatically. In consequence of the creasing, cotton dresses soon look untidy and so are out of favour. Tootal Broadhurst started a research department to tackle this problem, and put an old pupil of mine, Mr R. S. Willows, in charge. The problem took many years to solve and the investigation cost a good deal of money, but at last it was done and uncreasable cotton fabrics were obtained. The result is that, in spite of the depression in the cotton trade which has caused heavy falls in the price of most cotton shares, theirs are higher than they were before the depression. This is a very definite and direct demonstration of the advantages which a business may derive by having a research department as part of its organisation.

Changes such as I have mentioned did not come of themselves nor were they the work of one man. Huxley, Norman Lockyer, Lubbock, Michael Foster, Rücker and many others took an active part in the campaign, but no one was more successful or more active than Roscoe. Perhaps the most important factor of all was the interest aroused in science by great and revolutionary discoveries such as electric waves, argon, Röntgen rays and radium, which began in the early eighties and has continued ever since. This produced a crop of very substantial benefactions for the advancement of science, and also made it much easier than it would otherwise have been to persuade the Government to make large grants for the same purpose.

Roscoe's work as a chemist was in teaching and organisation rather than in discovery ; his original work was done quite early in his career, some of it in Bunsen's laboratory before he came to Owens College. He had the greatest veneration and love for Bunsen, and looked at chemistry from the same point of view. Bunsen was fond of saying that one new chemical fact, even an unimportant one, accurately determined, was worth a whole congress of discussion of matters of theory ; and this was Roscoe's opinion too.

It was amusing to hear Roscoe's treatment in his lectures of the idea of atoms ; when he was dealing with the law of chemical combination in multiple proportions he could not altogether evade it, but he obviously regarded it as somewhat flighty and Bohemian, and though it was evidently related to this very respectable and indispensable law, it was just as well not to parade the relationship. It was the irony of fate for Roscoe twenty years afterwards to discover, in the rooms of the Manchester Literary and Philosophical Society, Dalton's diary, which showed that, contrary to the view which he and others had held, it was the atomic theory that led to the law of chemical combination, and not that law to the atomic theory. Roscoe forgot that though a theory might be Bohemian it might be the parent of very respectable facts.

His courses of lectures were carefully arranged, were clear, and illustrated by plenty of experiments, which almost invariably came off. Once, when one did not, he rose to the occasion. He had told the class that when the contents of one test-tube were poured into those of another, the latter would turn blue. His assistant, Heywood, tried the experiment but something had gone amiss and the tube turned a bright red. He said, "Hold the tube up, Heywood, so that the class may see it. You see, gentlemen, what a beautiful and very peculiar blue it is." If you remembered his lectures you would have a very useful knowledge of the properties of the chemical elements. I did not for my own part find them very inspiring, though no doubt those who were already interested in chemistry found them very interesting. His *Lessons in Elementary Chemistry* contained a surprisingly large amount of information in a comparatively small space. It had an enormous circulation, over 200,000 copies having been sold by 1906.

SIR HENRY ROSCOE

I suppose for nearly forty years practically every English-speaking student of chemistry began by reading the *Lessons*.

By the time he left Owens its Chemical Department had become the best organised and the best equipped in the country, attended by over a hundred students. It was the first in England to have a Professor of Organic Chemistry, Schorlemmer, who was appointed in 1874 after having been a demonstrator for thirteen years. The department was a pioneer in yet another respect, for the Dalton Chemical Scholarship, established in 1856 in memory of Dalton, was to be awarded to the student in the chemical laboratory who did the best piece of original research. This was the first scholarship in England awarded for original research.

Besides his academic work, Roscoe did much by his own personal efforts to promote the application of science to industry. The manufacturers in Lancashire believed in him and were constantly coming to consult him as to the way they should get over difficulties which had cropped up in their works. He had a private laboratory at the College in which he made experiments to solve such problems. He was a very genial and hospitable man and immensely popular with his students, and often asked them to his house. I remember on one occasion when I was there, von Helmholtz, the great German physicist, and his daughter were staying in the house. It was the custom in those days for the hostess after dinner to ask some ladies who had been at the dinner to sing. Mrs Roscoe asked Miss Helmholtz, who said she would sing a certain song if someone would play the accompaniment. No one offered to do so, so she beckoned to her father and he came and sat down to the piano ; she hummed the tune, and he for a

29

few minutes kept beating it out with one finger, then he said he thought he could play it. He did so, and she sang her song.

When I went to the Owens College, I intended to take three years over the Engineering Course. Actually I remained at the College for five years. At the end of my second year my father died, when he was only thirty-nine, and the idea of my becoming an engineer had to be given up, as my mother could not afford the very large premium required for me to become an apprentice, which was the only avenue to the profession in those days. I had won some small scholarships which helped to pay the fees, and it was determined that I had better finish the three years' course, and obtain the certificate of engineering granted to those who had passed through the course to the satisfaction of the authorities. This I did, and got, besides the certificate, a scholarship in engineering and a prize for an essay on some engineering subject. This was the end of my career as an engineer, for Professor Barker advised me to stay at the College for another year and go on with mathematics and physics, and then try for an entrance scholarship at Trinity College, Cambridge. This was another turning-point in my career, for I had never thought of the possibility of going to Cambridge. The proposal was very attractive to me, for throughout the engineering course the subjects which interested me most, and in which I had done best, were mathematics and physics. I stayed on at the College, taking the higher classes in mathematics and physics and working at the physical laboratory. I tried in the spring of 1875 for an entrance scholarship at Trinity College, Cambridge : except for the examinations at the end of each course of lectures, this was the first time I had been in for an examination. My parents disliked examinations for young boys, and so, unlike the majority of my contemporaries at Owens, I had not taken any of the examinations of the London University which led to a degree. I was quite unsuccessful at my first attempt to get a scholarship at Trinity, and did not even qualify for an exhibition. The examiner wrote to Professor Barker and said that I should have done better if, instead of reading the higher subjects in mathematics which were not included in the examination, I had concentrated on getting a " thorough grounding " in the lower ones. This I think was true, but I am very glad that I had not done so. A " thorough training in elementary subjects " is often interpreted as reading the subjects included in the entrance scholarship examinations over and over again, and doing a great number of trivial examples in them. I had no idea of the extent to which this might be carried until I was chairman of a Royal Commission on Education, when evidence was brought before us that in some cases the boys, who were to be sent in to compete for entrance scholarships, did little in the two years before the examination but write out answers to papers set in previous examinations. Under this system the boys get more and more fed-up with mathematics the longer they are at it, while if they from time to time took up a new subject instead of revising an old one their interest in mathematics would continually increase, and they would come up to the university fresh and eager instead of stale and bored. I tried again the next year for an entrance scholarship and was successful, as I got a minor scholarship of £75 a year together with a subsizarship which entitled me to certain allowances. I think few can have owed more to scholarships than I do, for without them I could not have stayed at Owens or gone to Trinity. I must also express my gratitude for a scholarship I received from the Grocers' Company after I had been some time at Trinity ; it was of great service to me. Many years later they made me an Honorary Member of their Company—an honour which I value very highly.

Several of my contemporaries at Owens attained distinction in after-life. One of them, who was also my contemporary at Cambridge, became Archbishop of Perth in Western Australia. Six of them became Professors at either Oxford or Cambridge, one of them became Whip for the Liberal party, another President of the Institution of Civil Engineers. Perhaps the most widely known was G. R. Gissing, the novelist ; his novels are largely autobiographies, and the experience of the characters in his books when at College are his own experiences at Owens. For example, in *Born in Exile*, Whiteway College is Owens and one of its Professors is Adolphus Ward, who was Professor of English Literature and History in Gissing's time. Ward at once recognised Gissing's ability. I can remember at one of the annual meetings which were held at the end of each session, Ward, when announcing that Gissing had gained the prize for the English poem, said he had written a very rare thing, a prize poem that was real poetry. Ward took a great interest in Gissing, and was a good friend to him when he badly needed one. I did not know Gissing intimately, as we were not in the same year and did not meet at lectures, which were the usual places for meeting one's friends. As I remember him, he was pale, thin, listless, and looked and lived as if his means were small. After he left the College he lived by literature, but for many years he had a terrible struggle, hardly earning enough to keep body and soul together, and thinking when he could afford sixpence for his dinner it was a banquet. His life at this time is that described in his books, and especially in *New Grub Street*. Recognition was late in coming and many of his books yielded very little return. He did not receive much more than single copies of the first editions sold for thirty years after. Towards the end of his life he received a legacy sufficient to enable him to live in comfort. The difference this made on his outlook on life is shown in his writing. The *Private Papers of Henry Ryecroft*, written after, is full of the pleasures of life, while his earlier books had dealt with little but its miseries.

CHAPTER II

Undergraduate Days: Cambridge Then and Now

I CAME up to Trinity College in October 1876 and have "kept" every term since then, and been in residence for some part of each Long Vacation. As there was at the time no room in College when I came up, I went into lodgings at 16 Malcolm Street, intending to come into College when rooms were available. I found, however, the lodgings so comfortable and my landlady—Mrs Kemp, the widow of a Manciple at Trinity—so attentive, that I stayed in them for four years and did not come into College until I became a Fellow. I have never been able to remember, while I was working, to attend to a fire, nor work to any advantage when the room got cold, so I got on much better in lodgings where there was always someone about to look after the fire, than in rooms in College where the bedmaker would be away for the greater part of the day. My tutor was Mr J. M. Image : he was a classic, and I found this an advantage, for he let me choose the mathematical lectures I attended, whereas if he had been a mathematician he would have made me go to his own lectures. He was very helpful in other matters in which I required his assistance. The Master was Dr. W. H. Thompson, about whom I shall have something to say later, and the four tutors were Joseph Prior, H. M. Taylor, Coutts Trotter and Image. They were linked together in the song :

34

If little Joe Prior
Would only grow higher,
Not Trotter, nor Taylor, nor Image Esquire
Would be half such a man as little Joe Prior.

The point of "Image Esquire" was that Image, when he
met the freshmen at the beginning of the term in his classroom, informed them that he would not open letters
addressed —— Image, Esq. ; we must write—and here
he wrote on the blackboard—J. M. Image, Esq.

At that time it was not possible to take the Little-Go
before the end of the first term, and Greek was compulsory. I had not done any Greek, so I went to my tutor's
lectures on the Greek for the Little-Go. By the aid of
these, Bohn's translation of the set book, and a Greek
Grammar specially written for the Little-Go, and which
contained a long list of words which were irregular to
the point of impropriety in their behaviour, not one half
of which my classical friends had ever come across, I got
through. I did not spend much time over Greek, not
more than two hours a day for less than two months,
but the time so spent was utterly wasted : it was not
the slightest training in literature nor anything but a
useless strain on one's memory.

Besides attending lectures on Greek I also, like the great
majority of those aspiring to obtain a good place in the
Mathematical Tripos, " coached " with Routh, the most
famous of mathematical teachers. Routh's teaching was
not in the least like what is ordinarily understood by
" coaching ", it was in reality a series of exceedingly clear
and admirably arranged lectures, given to an audience
larger than that attending the lectures of many of the mathematical Professors or College Lecturers. I had heard so

much of Routh's teaching that I went to his class with great expectations. I confess at first I was somewhat disappointed : his lecture was quite clear, but there was nothing particularly novel or striking about what he said, and, taking any particular lecture, I had heard as good a one from other teachers. After a short time, however, I began to appreciate their merits, but to do this one must take into account what these lectures had to do. The mathematical tripos at the time I read with him included practically all the branches of pure and applied mathematics known at that time, and the examination was competitive —leaving out a subject would involve the risk of losing marks. The marks assigned to any subject might be a small fraction of the whole, so that several subjects might be omitted without altering the *class* in which the candidate was placed ; but it would alter materially his position in the class, and for men who hoped to be near the top of the tripos this was all-important. The best candidates read by far the greater part of the subjects, and they had to do this in three years and a term. A course of instruction which would give an adequate knowledge of such a large number of subjects in so short a time required a time-table that had been most carefully thought out and thoroughly tested, and everyone knew that he would get this if he went to Routh. He took his pupils in classes, and there were usually about ten in a class, and two such classes for each year. You were placed in one class or another at the beginning, according to the amount you had read and how you had done in the entrance scholarship examination before coming up, and the great majority of his pupils had taken this examination. Most of us remained throughout our reading with him in the class he had put us in at first. In his lectures he took us through the best textbook on the subject, the parts which the author had treated satisfactorily he just told us to read : when the book was obscure he made it plain ; when the proof of a theorem was longer than need be he gave us the shortest one ; when the author had put in something that was not important he told us not to read it ; when he had omitted something that was important he supplied the omissions. These diversions made the lectures more interesting and more easily remembered. His lectures on Rigid Dynamics, on which he had written the standard textbook which had been translated into several languages and which is still the standard work on the subject, were not so interesting as those on other subjects. He naturally had not so many opportunities for criticism.

The lectures were supplemented by manuscripts in his own handwriting on parts of the subjects which had not yet got into the textbooks and on which questions might possibly, though not probably, be set in the tripos. He did not touch on these in his lectures but referred us to the manuscripts, which were placed in a room next to his lecture-room at Peterhouse and which was open to all his pupils. Some copied them out, but as they were generally of considerable length and took a long time to copy, I contented myself with reading them. Another important part of his system of teaching was the weekly problem paper, which contained about a dozen problems taken from the different subjects set in the tripos ; one week we could take as much time as we pleased in solving the problems, the next we were expected to do them in three hours, the time allowed for such a paper in the tripos. We sent the papers in at the end of the week, and on the next Monday morning a complete solution of the paper in Routh's handwriting was placed in the pupils' room, together with a list of the marks each pupil had obtained. This introduced a sporting element, and made us take more trouble over them than we should otherwise have done.

Routh's system certainly succeeded in the object for which it was designed, that of training men to take high places in the tripos ; for in the thirty-three years from 1855 to 1888 in which it was in force, he had 27 Senior Wranglers and he taught 24 in 24 consecutive years. Results like these could not have been obtained unless he had been a born teacher, as he was, and had spent, as he had, time and labour in keeping his technique up to the mark. His lectures were given in a conversational way ; he was never eloquent, never humorous, but always clear. I do not think he would create enthusiasm for mathematics in those who had not it already, but he could better than any-one else give to students during their stay at Cambridge a sound and substantial knowledge in all the important branches of mathematics pure and applied. Until quite near the time when he gave up " coaching ", candidates for the Mathematical Tripos were expected to be acquainted with the whole of the wide range of pure and applied mathematics included in the examination. The range of reading was much wider and there was much less specialisa-tion under this system than in the ones which since 1882 have succeeded it, where the tripos is divided into two or more parts. The more elementary subjects are grouped in one part, and candidates are allowed to select a small number of the more advanced subjects for the other parts.

There is naturally, when we look back on our early days, a kind of glamour over many of our experiences and they look more attractive than they did at the time or than they would to an unbiassed judgment, but examinations are about the last things to which sentiment would cling, and I do not think it is prejudice which makes me prefer the system in vogue when I took my degree to those which succeeded it. I am glad that I came under the older system, for I probably read much more pure mathematics than I should have done if I had taken my degree a few years later. I have found this of great value (*c'est le premier pas qui coûte*), and it is a much less formidable task for the physicist, who finds that his researches require a knowledge of the highest parts of some branch of pure mathematics, to get this if he has already broken the ice, than if he has to start *ab initio*. Routh was such an interesting figure in the history of mathematics that I hope to be pardoned if I say a few more words about him. Example is better than precept, so he who teaches would-be Senior Wranglers ought to have been one himself. Routh fulfilled this condition, for he was Senior Wrangler in the year when Clerk Maxwell was second. Perhaps no other man has ever exerted so much influence on the teaching of mathematics ; for about half a century the vast majority of professors of mathematics in English, Scotch, Welsh and Colonial universities, and also the teachers of mathematics in the larger schools, had been pupils of his, and to a very large extent adopted his methods. In the textbooks of the time old pupils of Routh's would be continually meeting with passages which they recognised as echoes of what they had heard in his classroom or seen in his manuscripts.

He was the son of Sir Randolph Routh, K.C.B., and was born at Quebec. He came to England and studied mathematics under De Morgan at University College, London ; he entered at Peterhouse, Cambridge, in 1850. Clerk Maxwell also came up to that College in this year, but after one term migrated to Trinity. Routh, like

Maxwell, studied mathematics under Hopkins, the great
" coach" at that time, who had taught Stokes and William
Thomson, and scored 17 Senior Wranglers before he retired. Routh was Senior Wrangler and bracketed with
Clerk Maxwell for the Smith Prize. He began taking
pupils soon after taking his degree, and it was not long
before practically all the best men came to him for tuition.
He found time, in spite of all his private teaching, to write a
good many papers containing original researches in mathematics. These were all of high quality ; his most important work was, however, in the essay, " A Treatise on
the Stability of Motion, particularly Steady Motion ", with
which he won the Adams Prize in 1877. In this he introduced what he called " a modified Lagrangean Function "
which increased greatly the scope and power of Lagrange's
method, the most general and powerful of all dynamical
methods. Routh's work has been of fundamental importance in the application of dynamics to problems in physics.
He anticipated Sir William Thomson and von Helmholtz,
who had independently discovered the same theorem. He
was elected a Fellow of the Royal Society for his contribution to mathematics.

The regularity of Routh's life was almost incredible ;
his occupation during term time could be expressed as a
mathematical function of the time which had only one
solution. I believe one who had attended his lectures could
have told what he had been lecturing upon at a particular hour, and on a particular day, over a period of twenty-
five years. The fact that year after year he gave the same
lectures at the same time did not make him stale as it would
most people. He might, as far as one could judge from
his manner, have been delivering each lecture for the first
time.

His way of taking exercise was as regular as his lectures : every fine afternoon he started at the same time for a walk along the Trumpington Road ; went the same distance out, turned and came back. His regularity was not, as might perhaps have been expected, accompanied by formal and stereotyped manners ; these were very simple and kindly and we were all very fond of him. This was shown very markedly when in 1888 he gave up private tuition. His old pupils presented to Mrs Routh his portrait by Herkomer, and took the opportunity of expressing either by letter or by their presence at the meeting in Peterhouse, where the presentation was made by the late Lord Rayleigh, their gratitude to him for his teaching. I share to the full this feeling. It was a long-established custom for Routh's pupils to be photographed in a group in the term before the examination for the tripos. A collection of these would show what the great majority of the mathematicians of the last fifty years, and many other people who obtained eminence in other walks of life, looked like when they were undergraduates. My class at Routh's contained Joseph Larmor, Senior Wrangler and First Smith's Prizeman, who subsequently, like Newton and Stokes, became Lucasian Professor and representative of Cambridge University in Parliament, and was also, like Stokes, Secretary of the Royal Society; W. B. Allcock, who was Third Wrangler and became a Fellow of Emmanuel, and a most efficient and devoted teacher of mathematics in the College; and Homersham Cox, who was Fourth Wrangler and later a Fellow of Trinity College. Cox was one of the clearest-headed men I ever met but remarkably absent-minded. He became for a time a medical student : he did not take this very seriously. On one occasion, when in for one of the parts of the M.B. Examination, he found himself confronted with a paper on a subject which he had forgotten was included in the examination. He went to India as Professor of Mathematics in the University of Allahabad and was very successful in gaining the affections of his Indian students ; he had a very genuine sympathy with many of their opinions. Another member of Routh's class was E. J. C. Morton of St. John's College, who was President of the Union, an " Apostle ", and subsequently Member of Parliament. Martin Conway of Trinity, now Lord Conway, who has won great distinction in art, in politics and mountaineering, was a member for one year and then deserted mathematics for art. Very striking evidence of the importance attached to Routh's teaching is that so many undergraduates, many of them, like myself, poor men, were willing to pay his fees, amounting to £36 a year, to get it. This made mathematics a more expensive subject than classics or history, where private tuition was not nearly so general, and students were content with the lectures given by the Professors and College Lecturers. Though no doubt most of us went to Routh because we thought that we should get a higher place in the tripos by doing so, the teaching itself was well worth the extra expense.

Besides going to Routh, I went to lectures in Trinity College given by W. D. Niven on mathematical physics, mainly on Maxwell's treatise on *Electricity and Magnetism* which had then lately been published, and also lectures by J. W. L. Glaisher on pure mathematics. From both these I derived great benefit. Niven was not a fluent lecturer nor was his meaning always clear, but he was profoundly convinced of the importance of Maxwell's views and enthusiastic about them ; he managed to impart his enthusiasm to the class, and if we could not quite understand what he said about certain points, we were sure that these were important and that we must in some way or other get to understand them. This set us thinking about them and reading and re-reading Maxwell's book, which itself was not always clear. This was an excellent education and we got a much better grip of the subject, and greater interest in it, than we should have got if the question had seemed so clear to us in the lecture that we need not think further about it. The best teacher is not always the clearest lecturer but the one who is most successful in making his pupils think for themselves, and this Niven by his enthusiasm certainly did. After Maxwell's death Niven published for the University Press his collected papers, and prefaced them by an admirable biographical notice.

Niven was one of the best and kindest friends I ever had ; he was very kind to me from the time I came up as a freshman. He often asked me to go walks with him. I went very often to his rooms and, through him, I got to know many of the Fellows of the College.

He was an Aberdonian. He was bracketed with James Stuart as Third Wrangler in 1866 ; his brother Charles was Senior Wrangler in 1867 and afterwards Fellow of Trinity, and Professor of Natural Philosophy at Aberdeen and Fellow of the Royal Society ; and his brother James was bracketed Eighth Wrangler in 1874, afterwards Fellow of Queens' and Medical Officer of Health for Manchester. Sir William Niven was a Fellow of the Royal Society, and President of the Mathematical Society 1908–1909. He left Cambridge in 1882 and became Director of Studies at the Royal Naval College, Greenwich. He held this post until 1903, when he retired at the age of sixty, and received the K.C.B. for his official services. He died at Sidcup in 1917.

The lectures given by J. W. L. Glaisher on pure mathematics were the most interesting I ever attended on that subject; indeed he made me at one time quite enthusiastic about elliptic functions. The lectures were very clear and he covered a good deal of ground ; above all, they were never dull and were very human. If he was talking about a theorem discovered by X he would break out with " X was a great mathematician but he *was* a queer fish. Whenever he was introduced to a pretty girl he would, when he went home, write a sonnet about her and send it to her " ; or of Y, " He once stabbed a man and there was the dickens of a row ". This gave a certain liveliness to his lectures which was not conspicuous in those of some of his contemporaries. The dullness of some of these can hardly be imagined. One of them adopted a method which I have never seen before or since. He hardly spoke a word but wrote steadily on the blackboard ; when he had filled it he said, " Copy that ! " While we were doing this he was filling another board, and so the lecture went on. Glaisher's father was a well-known meteorologist who had gained great celebrity by making balloon ascents of record heights to take meteorological observations. Like his son he revelled in scientific societies, and his present to his son when he came of age was to pay the fees for life membership to the British Association, and to practically all the mathematical and astronomical societies whose membership could be obtained in this simple way. Glaisher was Second Wrangler in 1871, when John Hopkinson was Senior. He was then twenty-three years old. He was elected Fellow and Lecturer at Trinity

College in the same year. Before he took his degree he had had a paper published in the *Transactions* of the Royal Society, a most unusual thing for one so young. In 1871 he was put on a Committee of the British Association to report on mathematical tables. The other members of the Committee were Cayley, Stokes, William Thomson and H. J. S. Smith. This is very strong evidence of the high opinion which the leading mathematicians of the day held of his ability. He knew all the mathematicians of his time and had heard what they said about their predecessors, so that he knew a good deal more about mathematicians than their achievements in mathematics. During the seventies and eighties he was the most active promoter of research in pure mathematics in Cambridge ; he was the editor of the *Quarterly Journal of Mathematics*, in which much of this appeared. The conspicuous success in recent years of the Cambridge School of Pure Mathematics is due in no small degree to the spadework done by Glaisher more than fifty years ago.

In 1883 he accepted the offer of a tutorship in Trinity College. This was a most unfortunate decision. As a tutor he was subject to continual interruption which very seriously interfered with his mathemetical work, and when on Cayley's death in 1895 the Sadleirian Professorship in Pure Mathematics became vacant he was not elected. As his lectureship in mathematics at the College also ran out at about the same time, he badly wanted some other work to fill the gap and took to collecting china and pottery, and this became the chief interest of the last thirty years of his life. His original idea was to make a collection which would illustrate by early specimens the introduction of each step in the improvement of the technique of the potter's art. In this I understand he succeeded, but in doing this he had got bitten with the collector's mania, and became specially interested in " slip ware " and was anxious that his collection of this should be the best in the world. This is a very dangerous state of mind, for whenever an exceptionally fine specimen comes on the market you feel you must buy it or one of your rivals will get ahead of you. Glaisher spent a very large amount on his collection, considerably over £100,000 I believe, but he used to say that he had never regretted any of his extravagances, but had often been sorry for many of his economies which made him let a piece go which afterwards he was always longing to possess. He left his magnificent collection to the Fitzwilliam Museum : a beautifully illustrated catalogue made by Mr Rackham has just (1935) been issued. These will keep his memory green and give, as he would have wished, pleasure to many in the years to come. He was awarded the Sylvester Medal by the Royal, and the De Morgan Medal by the Mathematical Society. He had been President of many Societies, the Cambridge Philosophical, the Mathematical, the Astronomical, the Cambridge Antiquarian, and of the Cambridge University Bicycle Club, for he was in the early days of cycling an enthusiastic rider of the old " penny-farthing " machine. I think, however, the office which gave him the greatest pleasure was the Presidency of the Astronomical Club, which he held for thirty-five years. The Club is famous for good fellowship and good dinners. It is the custom of the Club for the President to propose the health of every guest at these dinners. By all accounts he did this with great success. Certainly the few speeches I heard were quite delightful, full of that humour, kind but discriminating, which was characteristic of the man. Though he was a bachelor, a don, and had lived in College rooms for more than sixty years, he was by no means a recluse ; in fact he was very fond of talking, and keenly interested in the affairs of his friends and acquaintances.

While I was an undergraduate I attended lectures given by Professors Cayley, Adams and Stokes. The first set of Cayley's lectures I attended were somewhat remarkable. I was the only undergraduate ; Glaisher and R. T. Wright, both Masters of Arts, made up the audience. Cayley did not use a blackboard, but sat at the end of a long narrow table and wrote with a quill pen on sheets of large foolscap paper. As the seats next the Professor were occupied by my seniors, I only saw the writing upside down. This, as may be imagined, made note-taking somewhat difficult. I should, however, be very sorry to have missed these lectures. It was a most interesting, valuable and educational experience to see Cayley solve a problem. He did not seem to trouble much about choosing the best method, but took the first that came to his mind. This led to analytical expressions which seemed hopelessly complicated and uncouth. Cayley, however, never seemed disconcerted but went steadily on, and in a few lines had changed the shapeless mass of symbols into beautifully symmetrical expressions, and the problem was solved. As a lesson in teaching one not to be afraid of a crowd of symbols, it was most valuable.

Professor Adams, whose lectures I also attended, never seemed to have any complicated collections of symbols to deal with ; he, like Kirchhoff, carried the feelings of an artist into his mathematics, and a demonstration had to be elegant as well as sound before he was satisfied. His lectures were wonderfully clear, and were read from beautifully written manuscript which he brought into the lecture-room in calico bags made by his wife. The lectures I attended were on his own researches on Lunar Theory. They had never been published and contained important additions to that theory. I think, however, reading a lecture deprives it of much of its charm : the reader is apt to be bored ; his mind is not so alert as if he had to frame his sentences as he went along, or rack his memory to reproduce them, and if he is lecturing to students he goes too quickly to allow them to take adequate notes. In particular, I think the dullest lectures are those read from the proof-sheets of a book which is passing through the press.

The lectures I enjoyed the most were those by Sir George Stokes on Light. For clearness of exposition, beauty and aptness of the experiments, I have never heard their equal. He had only the simplest apparatus at his command, no light but that of the sun, no assistant to help him. He prepared the experiments himself before the lecture and performed them himself in the lecture, and they always came off. His success was not obtained without much labour. His daughter, Mrs Humphry, in her biographical sketch prefixed to the *Memoirs and Scientific Correspondence of Sir G. G. Stokes*, edited by Sir Joseph Larmor, tells how preoccupied he was in the May Term not only with the experiments but also with the form of his lectures. His dependence on the sun obliged him to make hay while it shone, so that on bright days his lectures lasted far beyond the canonical hour. I remember on one occasion, long after my attendance at his lectures and when he was well over eighty, I met a batch of very hungry-looking students coming into the Cavendish

Laboratory one afternoon at half-past two. I asked them what they had been doing, and they said they had just come from Sir George Stokes' lecture which had commenced at twelve.

Like his predecessor in the Lucasian Chair, Sir Isaac Newton, he made his remarkable discoveries in optics working in his rooms in College with very simple apparatus. Though Stokes was such an accomplished lecturer, his power of keeping silent was equally remarkable. He, like Newton, represented Cambridge University in Parliament : he was member for four years and attended the House with great regularity, but never spoke. It is reported of Newton, who was member for two years, that the only time he addressed the House was to move that a window be opened, and Sir Joseph Larmor, another Lucasian Professor, was member for about eleven years and only spoke once, so that the Lucasian Professors cannot be accused of having wasted the time of the House. Stokes certainly had not much small-talk. A story is current in Cambridge that a visitor, who did not know that it was Lady Stokes she was speaking to, said, " There are two men in Cambridge whom it is positively painful to sit next at dinner : they never say a word ". Lady Stokes said, " Yes, George is one, but who is the other ? " He was, however, quite ready to talk on subjects on which he felt he had something to say. I once saw an amusing instance of this. He was sitting at lunch next a very charming American lady who started one subject after another without getting any reply but yes or no. At last in desperation, she asked him which did he like best, arithmetic or algebra ? The change was marvellous ; he became quite fluent and talked freely for the rest of the lunch.

He was quite ready to talk on scientific subjects, and a

talk with him left one wiser than before. He had a mind which, like a filter, seemed to clarify anything which passed through it, and one's ideas seemed always clearer and more definite after a talk with him about them. He was also very open-minded. When one presented to him a new idea, he did not at once say there could be nothing in it, but would listen quite patiently to anything one might say in its favour. He was, however, exceedingly cautious about coming to a conclusion. But, if he made any criticism, one was quite sure that it would be wise to reconsider most carefully the point he criticised before committing oneself to it. He had the Newtonian type of mind to a remarkable extent, but he was even more cautious than Newton. One might conceive his having written the bulk of Newton's *Optics* but not of his venturing to add the celebrated queries. It was very interesting to see Lord Kelvin and Stokes together. Lord Kelvin regarded Stokes as his teacher and always had the greatest respect for his opinion, and often said when he came to a difficulty on some mathematical or physical problem, "I must ask Stokes what he thinks about it ". The temperaments of the two were, however, very different. When Kelvin was speaking, Stokes would remain silent until Kelvin seemed at any rate to pause. On the other hand, when Stokes was speaking, Kelvin would butt in after almost every sentence with some idea which had just occurred to him, and which he could not suppress. I once saw a curious reversal of this. They had come together to the Cavendish Laboratory, and I was showing them some experiments I was engaged with at the time, on the electric discharge through gases. I happened to speak about atoms playing a part in one of the effects I was showing, when Kelvin said he did not believe in atoms but only in molecules. This was too much for

Stokes. He began at once to give a charmingly clear account of the reasons why atoms as well as molecules must exist. He was so much in earnest that Kelvin for once could not get a word in edgeways : as soon as he started to speak, Stokes raised his hand in a solemn way and, as it were, pushed Kelvin back into his seat.

Stokes was one of the Secretaries of the Royal Society for thirty-one years, and the work he did in this capacity had a very great effect on the progress of the physical sciences. He read all the mathematical and physical papers sent in to the Society with great care, suggesting improvement in the experiments, or in the arguments, or in the method of presentation. In addition to his work on the papers sent in to the Royal Society, he was in constant correspondence with many of the most active workers on physics, such as Crookes, Warren de la Rue, Smithells. These, as his correspondence and their acknowledgments show, often turned to him for advice. He was the guide as well as the teacher of his contemporaries. This work left him with but little time for his own researches. He had not had time to make any progress with the textbook on light which students had expected and hoped for for more than thirty years. Stokes, like Newton, was deeply interested in theology and spent much time in the latter part of his life in correspondence about his doctrine of " Conditional Immortality ". He held that only the souls of those who had been righteous in their life, or who had repented before their death, were immortal. The souls of the wicked vanished : there was a Heaven but no Hell. Some have regretted that he " wasted " on religious subjects so much time that might have been spent on scientific ones. It should be remembered, however, that his scale of values was probably different from that of his critics.

He was a deeply religious man, and to him as to many others the orthodox view about Hell and eternal torment seemed incompatible with the spirit of Christianity. He thought this was because " religious people insisted on a slavish literalism ", " framed a theory that the Bible must be interpreted in a way just like that in which a lawyer would interpret an Act of Parliament, stuck to the letter rather than attempted to catch the spirit ". He would have regarded time spent in clearing away obstacles to his own faith or in helping others to do the like, not as time wasted but as used to the best advantage.

" Conditional Immortality " did not get much support from Ecclesiastical Authorities and was, I believe, sometimes attacked from the University pulpit. Though a strong conservative in many things, he held liberal opinions about Church matters. He wished the compulsory signing of the Thirty-nine Articles to be abolished and the Athanasian Creed to be altered.

He was a pioneer in Cambridge in using a typewriter, much to the relief of his friends, as his handwriting was very shaky and somewhat difficult to read. It was, however, legibility itself compared with that of another Professor, whose letters took the spare time of many days to decipher.

My undergraduate days were very pleasant though uneventful. The most exciting incidents were perhaps the results of the annual College Examinations, and the migration for the Long Vacation term from my lodgings in Malcolm Street to rooms in the Great Court. In those days many more undergraduates came up for the Long Vacation and stayed up longer than they do now. Most men who were serious candidates for honours came up at the beginning of July, and stayed up to nearly the end of August. More than 200 came up then ; now there are not much more than half that number, and they do not stay up so long. It was very pleasant : we had more leisure as there were no College lectures, though the mathematicians went to Routh, and more opportunities of making friends. It was too, from the educational point of view, an important part of our stay in Cambridge. We had time to think, and to arrange and co-ordinate our knowledge. We got a better grip of our subject and were able to work out any new ideas we might have. An important event of the Long was the cricket match between members on the Foundation, *i.e.* the Fellows and the Scholars, and the rest. As at that time about half the members of the Cambridge eleven were Trinity men, there was generally a " blue " playing. There were, of course, some who never played cricket except at this match. It was an amusing match to watch. There was one classical Fellow of the College who was so anxious to make runs that he usually ran two or three of his side out in the attempt. The only fly in the ointment was that we had to " keep " three morning Chapels at 7.30 each week. This was more difficult than it looks, for sometimes though we got up early it was not quite early enough, and the Chapel doors were shut before we could reach them.

When I was in residence in the Long of 1879 the most severe thunderstorm I have ever seen struck Cambridge. It was on a Saturday night and Sunday morning : the storm had been going on for some time before I awoke, and when I looked out of the window the Great Court was a lake some inches deep ; torrents of rain were falling from the skies and pouring from the roofs into the Court. I looked at the river before going into Chapel on Sunday morning ; it was very high and rising rapidly, and when we came out it had overflowed its banks and the Paddocks on each side of the Avenue were flooded. It was many days before the water subsided from them, and there was fishing for small fish available for ardent anglers.

Another institution which was in force in my time, but which I am sorry to say has now disappeared, was " The Scholars' Table " ; one table at the dinner in Hall was reserved for scholars. They need not sit there unless they liked, but a large proportion of them did so. It was very enjoyable and gave excellent opportunities for making friends with men of other years and studying other subjects. T. E. Scrutton, a man in my own year who became Lord Justice Scrutton, dined regularly at this table. We all felt sure he would succeed at the Bar, he was so fond of arguing and so quick-witted, and so able. He got a First-Class in the Moral Science Tripos and was Senior in the Law Tripos. Later on he won the Yorke Prize on four occasions : two of the essays which won this prize, those on Charter Parties and on Copyright, developed into the standard books on these subjects. He was a prominent figure at the Union and became President, and was very successful in scoring off his opponents. When one of these had finished a long string of abuse by calling the other side " uncircumcised Philistines ", Scrutton at once jumped up and began, " My honourable and I suppose circumcised friend ".

Another man in the same year was H. H. West, a classical scholar, an Irishman, full of wit, immensely entertaining and a good actor. He gave on occasions representations of some of the figures in Smith's *Dictionary of Classical Antiquities* : his star turn was that of the man with the testudo. He took a leaf out of his table and, after taking off his clothes, put it on his back and gave a screamingly funny performance.

Another scholar of the same year was G. M. Edwards, a very good classic who was elected Fellow and Lecturer and afterwards Tutor of Sidney Sussex College. Another was Claud Melford Thompson, who was the first to win marks of distinction in both Physics and Chemistry in Part II of the Natural Sciences Tripos. He was for many years Professor of Chemistry in the University of Cardiff and took a prominent part in the administration of that university.

W. H. Whitfeld, a mathematical scholar in the same year, applied his mathematics to whist and succeeded " Cavendish " as Card Editor for the *Field*. He had an amazing card memory. I often played two or three rubbers with him after Hall, and I have known him, when I had been his partner, come round the next morning and tell me that if I had played another card instead of the one I had we should have won two more tricks. He took up a pack of cards and without any hesitation laid out the hand as it was when the wrong card was played. I of course did not remember more than a fraction of the cards, but the few I did remember agreed with his placing. He spent a great amount of time on the mathematics of whist, and by dealing out countless hands had accumulated a vast amount of material on how far the distribution of cards in actual play differed from that predicted by the Theory of Probability. By the advent of Bridge this had lost much of its interest and it was never, as far as I know, published.

G. W. Johnson, a scholar in the same year, took a Double First and went into the Civil Service and was for many years Chief Clerk in the Colonial Office. Another Double First had been obtained by E. V. Arnold, who was in the year above. He was bracketed Senior Classic and was also a Wrangler : he was a man of great energy and took a very active part in starting the *Cambridge Review*. He became Professor of Latin in the University of Bangor.

The Mathematical Tripos

The examination for the Mathematical Tripos when I sat for it in January 1880 was an arduous, anxious and a very uncomfortable experience. It was held in the depth of winter in the Senate House, a room in which there were no heating appliances of any kind. It certainly was horribly cold, though the ink did not freeze as it is reported to have once done. The examination was divided into two periods : the first lasted for four days. In the first three days we were examined on the elementary parts of geometry, conic sections, algebra and plane trigonometry, statics and dynamics, hydrostatics and optics, Newton and astronomy. Five papers were set on these subjects, each paper containing about twelve questions, each question consisting of a piece of book work and a rider which was a question, not supposed to be taken from a book, but one whose solution was closely connected with the piece of book work with which it was associated. In addition to these five book-work papers, there was a paper called the problem paper in which there was no book work. The questions were in themselves generally a little more difficult than the riders in the book-work papers, and, what was more important, you were deprived of the guidance given by their association with a particular piece of book work. In the papers set in the first three days you were not allowed to use

Photo: Hill & Saunders, Cambridge

DR. ROUTH AND PUPILS FOR THE MATHEMATICAL TRIPOS OF
JANUARY, 1880

Taken October Term, 1879.
Left to right: (*Back row*): J. W. Welsford, Joseph Larmor, J. Marshal, ——(?), J. J. Thomson, E. J.
C. Morton, F. F. Daldy.
(*Middle row*): T. Woodcock, Homersham Cox, ——(?), Dr. Routh, P. T. Wrigley, J. C. Watt.
(*Front row*): ——(?), A. McIntosh, W. B. Allcock.

the methods of the differential calculus, or of analytical geometry. The fourth day was a recent introduction dating from 1873. The two papers set on this day included easy questions on the higher parts of pure and applied mathematics, and also on the physical subjects—heat, electricity and magnetism—which had been introduced into the Tripos at the same time. An additional examiner was also appointed, specially qualified to deal with these subjects. The first of these examiners was Clerk Maxwell. On the fourth day the use of the differential calculus and analytical geometry was allowed. The papers in the morning were from 9 to 12, those in the afternoon from 1.30 to 4. At the end of the fourth day there was an interval of ten days in which the examiners drew up a list of those who, by their performance on the first three days, had acquitted themselves so as to deserve mathematical honours. These, and these only, could take the second part of the Tripos, which lasted for five days beginning on the Monday next but one after the beginning of the first four days. The marks allotted to the last five days were about six times those for the first four. The questions ranged over all branches of mathematics and there were two problem papers. The questions set on the first two days of this examination were on the parts of mathematics which had been included in the Tripos almost since its foundation, and were subject to the restriction that the book-work papers set in the first two days of the examination should not contain more questions than a well-prepared student might be expected to answer in the allotted time : this did not apply to the papers set after the second day. As each paper contained twelve questions and each question consisted of two parts, a piece of book work and a rider, this was pretty good going, but

I believe that a few men each year did get a large percentage of the maximum marks obtainable on these papers. They knew the shortest proof for each piece of book work that was set; they had written it out over and over again; they had not to think about it, but merely write as fast as they could. The riders were a greater difficulty, but in the textbooks numerous " examples " were given for each piece of book work and they had probably done something very similar to the rider set in the paper. The details of the question, *e.g.* the algebraical work, was new, and it was of great importance that the student should make no slips in this; an error in arithmetic, or a wrong sign in a piece of algebra, would involve going through the work again, and a loss of time when there was no time to lose. Accuracy in manipulation was perhaps the most important condition in this part of the examination and the most difficult to impart : to err is human even in mathematics.

Another quality which played a great part was concentration on the question in hand, and ability to get quickly into stride for another question as soon as one had finished with the old. These qualities, having one's knowledge at one's finger-ends, concentration, accuracy and mobility owed their importance to the examination being competitive, to there being an order of merit, to our having to gallop all the way to have a chance of winning. These qualities are just those which are of most importance at the Bar, and made the old Mathematical Tripos an excellent training for that profession. It is significant, I think, that the last Tripos to give up arranging the list in order of merit was the Law Tripos, which retained it until 1923, forty years after it had been dropped in the mathematical. It is somewhat humiliating to compare the rate at which one could do mathematical problems when in for the Tripos, with the rate one can attain late in life.

In the last three days of the Tripos, when no pretence was made that the questions could be done by well-prepared students in the allotted time, a much smaller percentage of marks was the rule. The book work in the higher subjects took much longer, and the number of questions was only about two less than in the other parts ; indeed it would have been good going to write out the book work alone. The riders, too, took in general longer to write out, and since sometimes not more than one examiner was an expert in some of the new subjects, his papers were not subject to effective criticism from his colleagues, so that sometimes the papers were absurdly difficult. I have heard of cases where not a single question in a long paper was completely done by any of the candidates. This led to there being great gaps between the marks obtained by candidates coming next in order in the list. There was one famous case when the Senior Wrangler's marks were considerably more than twice those of the Second Wrangler, who became a very distinguished mathematician and a Professor in the University. From my experience as an examiner in both mathematics and physics, I have come to the conclusion that examinations in the highest parts of these subjects are unsatisfactory. In mathematics the pieces of book work are in general very long, and if the candidate makes a slip in the analysis and does not get the result asked for, it may take him a long time to find out the error and correct it. I have known more than one case where a candidate, who later proved to be a very able mathematician, sent up a paper in which not a single question was completely solved. When preparing for an examination one has to get up the subject in much greater detail than is necessary for understanding its scope, and the general methods used to obtain the results. Memory is burdened with a mass of detail which will very soon be forgotten. All that one need remember is where to look for them when wanted. The same considerations apply to examinations on the most recent discoveries in physics. In each case I think the student would benefit if, instead of spending so much time in getting his knowledge up to examination pitch, he spent some time in attempting a piece of research.

Again, the questions set in mathematical physics often stop where the real interest begins. The question asked is often to find a relation between a number of mathematical symbols representing various physical quantities ; nothing is asked as to what are the physical consequences resulting from this relation. This, however, is just the thing which is of interest to the physicist ; it is as if he received a message in cypher and made no attempt to decode it.

The Tripos under the regulation I have described was, in my opinion, a very good examination for the better men. It was, however, a very bad one for the majority who had not exceptional mathematical ability. Many of these men could cope with the more elementary subjects and benefit by studying them, but with these they had shot their bolt ; they found the higher subjects beyond them and the time they spent over them wasted. Under the old regulations, if these men were to get any credit for the mathematics they did know, they must wait for the Tripos at the end of their tenth term, and for the last three or even more terms they were only marking time, not gaining any additional knowledge or any mental training.

Under the new system the men can be examined in elementary mathematics at the end of their first or second year, and can thus take up a new subject, such as engineering, economics or science, in which their mathematics may prove of considerable assistance. Whether or not the changes that have been made in the Mathematical Tripos have been instrumental in promoting the progress of mathematical science in this country is a question which it is very difficult to answer. There is no doubt that the number of mathematical papers published in England has increased very greatly, and so has the number of people engaged in research in mathematics, but this in my opinion is largely due to the great change which has come over the feeling in the country with regard to research, rather than to the changes in the Tripos itself. In the days of the old Tripos very few rewards were given for research. Fellowships, except at Trinity College, were given for success in examinations and not for research. Scholarships stopped as soon as the scholar had taken his degree. When a man applied for a teaching post the qualification which carried most weight was that of being a good teacher. No one asked whether he was a good researcher, or whether he had done any research at all. There were no rewards for research except the small number of professorships in the few universities which then existed. Now all this has been changed ; the country has got the research habit, research is the fashion in all subjects. Candidates for Fellowships would have little chance of election unless they submitted a dissertation containing an account of a piece of original research. For all scientific posts the predominant qualification is a successful piece of research. The student who, after taking his degree, wishes to research, may now get very substantial grants of money to support him while researching. These may come from his College in the form of research scholarships, from the university which awards studentships for research, from outside bodies like the Board of Invention and Research, which has made a great number of substantial grants to enable students to continue their researches ; from the 1851 Commission and some of the great City Companies, to mention only the most prominent. There are thus many more students in the university engaged in researches, each of which ought to, and many do, result in the production of a paper worthy of publication in a scientific magazine or the Transactions of some scientific society. The great majority of these would not have been written unless the authors had received inducements and assistance which hardly existed fifty years ago. Then there were comparatively few men engaged in research, and those who were, did it not because public opinion expected it of them, but because of their enthusiasm for science and the joy they felt in discovery. They were exceptional men. Men of this type cannot be produced by giving scholarships or even by particular schemes of education. They are born, not made ; and as experience has shown, they seem able to get to the front in spite of all obstacles. The " mute inglorious Milton " is probably a myth. Dr. Rashdall (*Universities of Europe in the Middle Ages*, vol. ii. pp. 2, 602) says that the means of education in the Middle Ages were more widely diffused than is generally supposed, and that " except in remote and thinly populated regions, a boy would never have to go very far from home to find a regular Grammar School ". Thus any boy of exceptional ability might be expected to have come under the notice of someone who, like the old Scotch dominies, would set to work to raise funds enough to enable him to go to a university. Increase in the sum spent on education, improved methods of teaching, the provision of scholarships to the universities and the like, cannot be expected to increase to any great extent the number of quite exceptional men. They can, however, and do, increase the number of able men available for the service of the State, for our industries and for the promotion of science. They may be compared with improvements in telescopes which, while they do not reveal new stars of the first magnitude, which can be seen without them, do bring to light multitudes of new worlds.

I took the Tripos Examination in January 1880. The thing about it which remains most clearly in my memory is that I suffered from a bad attack of insomnia during the last five days, and got very little sleep. Insomnia is even more unpleasant in Cambridge than in most other places, for since several clocks chime each quarter of an hour you know exactly how much sleep you have lost, and this makes you lose more. Though the nights were very unpleasant I do not think they had much effect on my work in the examination, for when I got started on a paper the feeling of fatigue passed away. I had a shampoo between the morning and afternoon papers and got much sympathy from the barber, who said he was glad he was not in my shoes. I came out Second Wrangler; Larmor was Senior. I was very well satisfied with the result, as I was quite as high up as I ever expected to be. I had a return of insomnia in the Smith's Prize Examination and actually had a short doze in the afternoon paper at the Observatory. This may, however, have been due to the excellent lunch which Professor and Mrs Adams provided for us.

Undergraduate Life Then and Now

I ceased to be an undergraduate in 1880. Looking back on my undergraduate days and comparing undergraduate life then with what it is now, I see of course many important changes but not, I think, such important ones as have occurred in many other walks of life. They are certainly far less fundamental than those in the lives of the dons, for hardly any of these were married when I was an undergraduate, while now but few are single. The advent of students from Newnham and Girton has made some difference. We did not sit side by side with ladies during lectures, nor treat them to chocolate and cakes in the intervals between them, nor, except in May week, invite them to our rooms. These, however, are trifles compared with matrimony. Perhaps the most striking difference is in amusements and games. In those days there were no theatres open in term time. The Theatre Royal in Barnwell, a suburb of Cambridge, existed, but was only open during the Long Vacation. We find from *Gunning's Reminiscences* that it was regularly attended at the end of the eighteenth century during Stourbridge Fair, then one of the largest fairs in the country, by a group of Shakespearian critics headed by Dr. Farmer, the Master of Emmanuel College. It had in my time grievously fallen from that high estate, and some of the interpretations of Shakespeare's meaning would not have met with his approval.

One year when we came up at the beginning of the October term, we found posters all over the town announcing that the Theatre Royal, Barnwell, would reopen on a certain day under entirely new management ; when the audience arrived they found that instead of a play they were offered a prayer meeting. The theatre had been bought by the Hon. Ian Keith Falconer, who subsequently became Lord Almoner's Professor of Arabic, and who was an ardent supporter of religious missions. He was also the champion bicyclist of his time. The theatre was renovated and revived a few years ago as the " Festival ". It produced many interesting plays which could rarely be seen elsewhere, and became an important feature in the social life of the University.

The only plays to be seen at Cambridge in my days were one in the Michaelmas Term and another in the May Term, given by the A.D.C., a dramatic club founded by Burnand. These reached a very high standard. There was generally among the older members of the University someone who, like J. W. Clark in my time, took great interest in the theatre, and who spent a great deal of time and trouble over the rehearsals. All the women's parts were taken by men, and I think the best Lady Teazle I ever saw was Mr Manners, a Jesus man. Some of the undergraduates who took part in these plays subsequently went on the stage, and one or two of them attained considerable success. The photographs in the rooms of the A.D.C. are very interesting. You see future Bishops as singing chambermaids and Cabinet Ministers as butlers. Things are changed now. There is a large theatre and another being built ; there are five or six cinemas ; there are several new dramatic clubs such as the Marlowe Society and the Footlights, and every few years there is a Greek play. Thus each year several plays are produced, and a great deal of time and trouble is spent over them, and the work of the undergraduates taking part in them seriously interrupted. There was always much good music to be heard in Cambridge, but now the Cambridge musicians are much more ambitious, and produce operas such as Purcell's *Faery Queen* with large orchestras, and which involve a great sacrifice of time by those taking part in them. In addition to music there is dancing. This in my time was practically confined to a ball in the May Week ; now since there are so many more ladies resident in Cambridge, dances are going on all through term time, and besides private dances, there are dancing clubs, and dances for every conceivable charitable purpose. The result is that an undergraduate who is good at acting, singing or dancing gets little time for his studies in term time and has to rely on the vacations for his reading for his examinations.

In those days, roughly speaking, the only games played by undergraduates were cricket and football ; lawn tennis had only just been introduced, and squash racquets was a long way ahead. There certainly was a golf-links at Coldham Common quite close to the town, but these I think were the most depressing and uninviting links I ever saw. So few played that a friend of mine who went there one afternoon and was knocking a ball about with a club, was seized upon by the Secretary and asked to play against Oxford the next day. A few played fives, or real tennis, at the courts in Burrell's Walk or Tennis Court Road. The ordinary reading man who was not particularly good at games had not much chance of playing either cricket or football, for there were enough men who had got their " colours " at school to fill up the College teams—there was no room for the " rabbits ". A fair number of freshmen joined their College boat club, where they found a good many companions who, like themselves, knew very little about rowing. They went regularly in their first term down to the river to be tubbed. Rowing is not a very satisfactory way for the ordinary reading man to get his exercise. If he does not take it seriously enough to get into a College eight, he does not get enough exercise, while if he does get in he gets too much, and does not feel inclined for work in the evening. In those days the great majority of the reading men got their exercise by taking walks. Between 2 and 4 in the afternoon, streams of undergraduates, two and two, might be seen on all the roads within three or four miles of Cambridge, taking the Grantchester, Madingley, Coton, Cherryhinton or Shelford " grinds ". On Sundays many went further afield and walked for five or six hours. Now there are many games which, like lawn tennis, golf or squash racquets, only require two players and for which it is generally possible to find an opponent nearly as bad as yourself, so that you need not feel you are spoiling any-one's game. Even in football, hockey or cricket, there are many clubs which cater especially for poor players, so that now very few undergraduates take walks. I found these walks very pleasant ; the scenery round Cambridge is much better than most people imagine. Besides, if you keep your eyes open there are beautiful things to be seen in any walk in the country, and even in one in a town. I think it is a great pity that children both at home and at school are not encouraged to look out for beauty around them. If they once got the habit it would a be continual source of pleasure to them. Most people nowadays hardly recognise beauty unless it is pointed out to them, and will go miles to see something which figures on a picture postcard, while more beautiful things are to be found close by them. The " model villages ", the " fairy glens ", the " pretty corners ", the " lover's leaps " of the picture postcards, might all go without diminishing by more than an infinitesimal fraction the beauty of the country. The walks and the long talks which accompanied them, were an excellent means of making friendships more intimate. The *solitude à deux* is, I think, much better for this purpose than playing games, or going to meetings to listen to essays or discussions.

One very great friend of mine with whom I often took such walks was J. C. Watt, who had come up to Trinity from Glasgow University, where he had been a pupil of Sir William Thomson. After staying two years at Trinity he migrated to Jesus College ; he was elected to a Fellowship there in 1881, and to a lectureship in 1883. He never married, but lived in College until his health broke down a few years before his death in 1931.

He possessed to a degree I have never seen surpassed the power of winning the confidence of undergraduates.

He welcomed them to his rooms, took great interest in what they were doing, and knew all about their successes in games, for he rarely missed an important cricket or football match. This put them at their ease, and helped to reveal to them his kindliness, which was the secret of his success.

They felt that they had in him a real friend to whom they could go if they got into difficulties, even when these were due to their own silliness. They did so, and disclosed what they had done with a frankness much greater than most of them could have done to their own fathers. He began by telling them very frankly what he thought of their conduct, and then set to work to try to help them.

Many Jesus men who were up between 1883 and 1928

owe much to the help they got from " Tommy " Watt,
as he was always called.

A great change in the relative popularity of different
games has occurred since I was an undergraduate. Then
the Rugby football matches were played on Parker's
Piece ; there was no charge for admission, and the attendance was not a tenth of what it is now when the University has a fine ground of its own, where the gate money
puts the finances of the Rugby Club into such a satisfactory
position that it not only makes generous donations to
support clubs in other sports, but has actually founded a
lectureship in Classics in the University. On the other
hand Fenners, the University cricket-ground, is now almost
deserted, unless the match is with the Australians or teams
from India or the Dominions. Then there would be a
ring three or four deep round the ground at every match
in the May Term. The pavilion was crowded with dons ;
you were pretty sure to see there Munro, the great classic ;
Aldis Wright, the Hebrew and Shakespearian scholar ;
James Porter, the Master of Peterhouse, accompanied by
his dog, for which he had paid the fee for life membership
of the Cricket Club, so as to quiet his conscience about
breaking the rule that dogs were not admitted to the
ground ; Ward of Magdalene, a very stout man who was
distinguished from his brother, one of the protagonists of
the Oxford Movement, by being called the real presence,
and his brother the ideal presence.

Munro took great interest in cricket and, as he was a
Trinity man, worked to have that College well represented
in the Oxford and Cambridge cricket match. I remember
one morning going into the Combination Room in
College to read the newspapers ; the only person in the
room was Munro, who was reading *The Times*. I had

not met him before and he looked very disappointed when
he saw me.   It was evident there was something he wanted
to talk about very badly.   I sat down and began reading
the paper.   After a minute or two Munro could stand it
no longer.   He dashed *The Times* with quite a bang on
the floor and said, " What do you think that fellow
Verrall's done ? " (A. W. Verrall was a distinguished
lecturer in Classics in the College.)   I said I did not know.
" Well," he said, " he's ploughed Foley for entrance to
the College (Foley was a boy who had made a century
in the Eton and Harrow match), and how do you think
Verrall translates ? "   Here he went off with a piece of
Greek, and gave Verrall's translation.   I tried to look as
shocked as I could, and he went on, " I don't know what
this College is coming to, when a man who makes
mistakes like these in a straightforward piece of translation
is allowed to plough a boy like Foley ".

Cambridge cricket was then at its zenith.   Edward
and Alfred Lyttelton, D. Q. and A. G. Steel, A. P. Lucas
and Ivo Bligh all played for Cambridge between 1876 and
1880.   Edward Lyttelton was captain the year he took the
Classical Tripos, and used to bring his books to Fenners
and read while his side was batting.   The most decisive of
the few defeats suffered by the first Australian team to
visit the country was that inflicted in 1878 by Cambridge
University, and on their second visit in 1882 they were
again defeated by Cambridge.

In those days " blues " were given only to those who
represented Cambridge in the boat race or the cricket
match : no blues were given for football.   Oxford,
however, gave " Rugger " blues in 1884.   The Cambridge
Rugby team was naturally up in arms, and said that if
blues were not given to them they would take them.

This brought all the wigs on the green, and in 1885 a meeting, to which all members of the University were admitted, was held in the rooms of the "Union", the largest undergraduate club in Cambridge, to ascertain the feeling of the University. My old friend Richard Threlfall, of whom I shall have more to say later, who had twice played in the Rugby team against Oxford, was selected to plead the case for the "blue" for the Rugby team. I went with him to the meeting, and though we arrived a few minutes before it was timed to begin, the room was so packed that he could not get through the crowd at the back of the hall, and he had to be held up by a few of his friends, of whom I am proud to say I was one, while he made his speech. The speakers before him were against the proposal, and delivered carefully prepared speeches which smelt very strongly of the lamp and left the audience quite cold ; when he got up and jerked out from his uncomfortable stance one short sentence after another full of good sense, good humour and good jokes, he soon had the house rocking with laughter, and put the issue beyond doubt. I never heard a speech which had so much influence upon the division.

The views held by undergraduates as to what is good form change almost as frequently as the fashion of ladies' dresses. In my time and for many years afterwards it was considered the height of indecency to carry an umbrella when wearing cap and gown, indeed our tutors warned us against it at their first interview. I have lived to see a time when it was considered to be indecent to be without an umbrella when you were out of doors, whatever other garments you might lack. The correct way of wearing it was not to have it rolled up neatly, but baggy like Mrs Gamp's : it was not used as a walking-stick but carried gingerly in front, without touching the ground. Now again the umbrella has, for a time at least, retired to the background.

Since the advent of Newnham and Girton students, undergraduates' dress has been influenced by the necessity of preserving an adequate distinction between male and female attire and behaviour. In the days when the " Eton crop " was in fashion with Girton and Newnham students, a Trinity undergraduate with rather long hair called on his tutor. As he was going away the tutor said : " If I were you I would get my hair cut ". " Oh, sir," said the undergraduate, " I should not like to do that, short hair is so effeminate." Since cigarette-smoking has become so usual with women it has lost favour with undergraduates. The proportion of non-smokers to smokers among Trinity undergraduates is far greater than it used to be and is, I think, greater than that at Newnham or Girton. If the fashion for women to wear trousers extends to these Colleges our undergraduates will be driven to kilts.

The modern undergraduate is more businesslike than we used to be, and more careful to get good value for his money. My tailor tells me that they now nearly always ask the price of a suit before ordering it, while fifty years ago this was very unusual. Ready-money transactions are much more usual, surprisingly so, considering the readiness with which tradesmen give credit to undergraduates. A century ago Cambridge tailors had a peculiar system of insurance against losses from unpaid bills. My father-in-law, Sir George Paget, told me that a College friend of his went down without paying his tailor's bill. He got a post, and in about four years had made enough money to enable him to pay his debts. He came up to Cambridge to do so, went to his tailor and said he wanted to pay his bill. The tailor said, " Sir, I could not think of taking the money ; we make it a rule when a gentleman has been down for three years without paying his bill, to distribute the amount he owes us among the bills of our other customers ". I got an illustration of the change which had come about, in a conversation I had with a bootmaker early in this century. I said I hoped trade was good. " No, sir," he said, " it is not. Things are very different now from what they were when you were an undergraduate. You will hardly believe me, sir, when I tell you that I have not met what I should call a really extravagant gentleman for more than three years." " Well, Mr ——," I said, " what would you call an extravagant gentleman ? " " Sir," he said, " I should call a gentleman extravagant if he had more than two pairs of boots a week."

When I was an undergraduate the great majority of Cambridge men, after taking their degree, went into one or other of the professions, the Church, the Bar, Civil Service, Medicine or schoolmastering. Very few went into business, and those who did generally went into firms with which they had some family connection ; this is now changed, and more go into business than to the professions. This is because those at the head of large business concerns attach much greater importance than they used to do to a university education, and are willing to give, without the payment of a premium, a trial to those who have done well at Cambridge. In my time Cambridge men, after taking their degree, had to find a post for themselves ; there is now an important branch of the University called the Appointments Board, which has offices and a permanent staff in Cambridge, and has the advice of a committee consisting of some of the men most influential in our industries, finance and education. An old colleague of mine at the Cavendish Laboratory, now Sir Napier Shaw, F.R.S., had a great deal to do with its formation. Under the guidance of its first secretary, H. A. Roberts, it gained the confidence of the employers ; he took great pains to get reliable information about the applicants for employment ; he acted more like a judge than an advocate, and when he recommended a man, would mention his weaknesses as well as his merits. The result of this was that from small beginnings the Department rapidly increased ; in the last Academical year 1935-36 it found posts for 596 men. These posts are of all kinds, financial, engineering, chemical, professorial, scholastic, secretarial, and most of our undergraduates who want employment after they leave the University send in their names to the Appointments Board.

The increase in the number of provincial universities and schools of technology as well as in that of the professors, readers and lecturers in the older universities has provided a much larger number of academic posts. The attraction of these has drawn in many who would otherwise have become schoolmasters, especially those who have taken high honours.

The greatest change, however, has been in the number of men who take Holy Orders. Of the 909 undergraduates who entered Trinity College in the years 1920-24 inclusive, and whose names are on the books of the College, only 24 have done so, and all the indications point to a further decrease in this proportion.

CHAPTER III

Cambridge, 1879–1884

AFTER taking my degree I began the preparation of a dissertation to be sent in for the Fellowship Examination. The idea on which this was based had occurred to me before I came to Cambridge (see page 21), but I had not, while preparing for the Tripos, had time to develop it. It had to do with the nature of energy. According to the views then prevalent there were many different forms of energy. There was kinetic energy, that due to the motion of mass ; potential energy, that due to the position of a body amid various surroundings ; thermal energy, that possessed by a hot body in virtue of its temperature ; electromagnetic energy, that in electric and magnetic fields ; energy of chemical separation, that liberated when two different substances, e.g. coal and oxygen, combine and give out heat. Energy of one kind could be converted into that of another but there was always a fixed rate of exchange, e.g. if a certain amount of kinetic energy were converted by friction into heat, the thermal energy developed would be strictly proportional to the amount of kinetic energy lost ; or, comparing different kinds of energy with different currencies, if all kinds were estimated on a gold standard the principle of the conservation of energy asserts that their total amount could not be altered by any physical process. The passage of energy from one form to another was called the transformation of energy. The view that energy itself could be of quite different kinds had always seemed to me to lead to great difficulties when one tried to form a mental picture of what was going on when one kind of energy was transformed into another, and that a much simpler view was to suppose that energy was all of one kind and that the transformation of energy could be more appropriately described as a *transference* of energy from one system to another, the physical effects produced by the energy depending on the nature of the system in which it finds a home. My dissertation was of the view that all energy was kinetic energy. Very general methods for dealing with systems possessing only kinetic energy have been developed by Lagrange and Hamilton and are expressed by the Hamiltonian and Lagrangian Equations ; my dissertation was the application of these, and especially the Lagrangian ones, to various problems in physics and chemistry. This led to interesting relations between various physical effects which must be true whatever may be the kind of mechanism which produces the effects. For example, if magnetising a bar of iron alters its length, altering its length will alter its magnetism, and if the amount of one of these effects is known, that of the other can be calculated. Again, if a property varies with the temperature, *e.g.* if the stiffness of a spring depends on its temperature, pulling out the spring will alter its temperature and the alteration in temperature will be such as to make it more difficult to pull it out. Thus if the spring is stiffer when it is cold than when it is hot, pulling out the spring will cool it, while if it is stiffer when hot than when cold, pulling it out will heat it. The general principle is that the alteration in temperature produced by any change in the system is that which will increase the resistance to that change. This principle was published independently and almost simultaneously by Le Chatelier. The dissertation was published in an expanded form in two papers in the *Transactions* of the Royal Society, and was the foundation of my book *Applications of Dynamics to Physics and Chemistry*.

In those days candidates in mathematics for a Fellowship at Trinity, besides sending in a dissertation, had to take an examination in which there were two mathematical papers and another paper on a subject (not mathematical) chosen by themselves. I had a peculiar experience with regard to the subject I chose for this paper. My first choice was Marshall's *Principles of Political Economy*, as I thought that perhaps my mathematics might help in that; but when I found that on almost every page general laws in large print were stated on what seemed to me very insufficient evidence, I thought I had better try something else, so I changed over to Kant. I don't suppose I really understood it, but I enjoyed reading it and found it much more satisfactory than the *Political Economy*. I was fortunate enough to be elected to a Fellowship at my first attempt, to the great surprise of my tutor. In the interval between the end of the examination and the announcement of the result I met him in the Great Court and he said, " What are you doing here at this time of the year ? " I said I had just been in for the Fellowship Examination. He said, " That is just like you, Thomson, never asking my advice. If you had come to me in the May Term I could have told you that there are two Senior Classics having their second try and a very strong Science candidate having his third and last, and there is not the slightest chance of their taking a candidate at his first try. You could then have enjoyed yourself in the Long Vacation instead of wasting your time over the Fellowship." I was, however, lucky enough to be elected.

In those days a Fellow did not enter into all his privileges and emoluments until he became a Master of Arts ; he could not dine at the High Table and only received two thirds of a Fellow's dividend. To compensate for this the time he was a B.A. did not count in the seven years which was the tenure of a Fellowship.

The Bachelors of Arts, whether Fellows or not, dined together at a separate table called the "Bachelors' Table". This was managed by the Senior Bachelor, who was the Bachelor of highest seniority, a Bachelor Fellow being regarded as senior to all Bachelors who were not Fellows. The office involved a good deal of work, for the Senior Bachelor had each evening, with the assistance of the Head Waiter, to order the dinner for the next day. It involved also some financial responsibility, for he was responsible for the payment to the College kitchens of the expenses of the dinners ; he was credited with a fixed sum for each dinner eaten by a Bachelor ; but if the kitchen bill at the end of the term exceeded that allowance, he had to pay the difference. As I had got my Fellowship very early, I was Senior Bachelor for nearly two years, and on the whole found the work agreeable. It was difficult, however, for the Senior Bachelor to dine out as frequently as he might wish to do, for if he were absent the dinner was ordered by the Bachelor next in seniority, and as he had no financial responsibility he often ordered the things he liked best with little regard to their cost. One night's absence might require economies for several days to get this account straight. Though the Bachelors' Table still exists, the arrangements are quite different, and a Fellow dines at the High Table as soon as he is elected.

Soon after taking my degree, I lectured for three hours on three mornings of the week at Cavendish College, a college which had been founded to provide a place of residence for undergraduates younger than those usually admitted to the older Colleges. It was thought that perhaps one of the reasons why so few undergraduates went into business was because business men thought a start should be made earlier than the usual age for leaving Cambridge. The College, however, was not a success, and after a few years was given up. It is now under the name of Homerton College, a College for the training of women teachers.

I also took some private pupils; among the first was Austen Chamberlain, now Sir Austen Chamberlain, K.G., who came because, under the regulations then in force, no one could take a Tripos in any subject unless he passed a preliminary examination in mathematics. I remember his father coming to Cambridge about this time and asking how his son was getting on. When he was told that he was doing very well and was certain to pass, he said he was not surprised, for he had never had but one fault to find with Austen, and that was he was such an awful Radical. Another of my pupils at this time was Eldon Gorst, who subsequently became High Commissioner for Egypt; he was a candidate for the Mathematical Tripos and took a First Class in it. His motto was " Thorough ". When he came back at the beginning of the term in which his Tripos took place, he mumbled so that it was only with the greatest difficulty I could make out what he was saying. I asked what was the matter, and he said that all his life he had suffered from toothache and was determined that he would not be hampered by it in the Tripos, and so had had every tooth in his head taken out; as his gums were not yet healed, the plate could not be put in, and so he could not articulate properly.

I think it helps any teacher to have had some experience of teaching pupils one by one ; he realises what are the difficulties they meet with much more vividly than if he lectures to a class and only comes into contact with them through the written answers they send in to questions set in the lecture. I learnt, too, from my experience with these pupils, how greatly in most cases the concrete exceeds the abstract in the power of stimulating the mind and facilitating thought and work. One of my pupils came at the beginning of his third year with a very bad record. He was said to be idle, to take no interest in his work, and to have very little chance of getting through the Tripos. At first I agreed with this estimate, but we plodded on until we came to the subject of collisions between elastic spheres. I knew he was fond of billiards, and so I pointed out to him that the rules he used for playing certain shots at billiards followed at once from the mathematics. The result was marvellous. He had never before conceived that there was anything in mathematics that could interest any reasonable being. He now respected it, and began to work like a nigger : he was quite intelligent, and in the few months that were left before the Tripos he learnt enough to take quite a respectable place among the Senior Optimes. If it had not been for the billiards he would have been ploughed.

I was elected to an Assistant Lectureship in mathematics at Trinity College about two years after taking my degree. At this time the College were taking steps to remove the anomaly—to call it by no harsher name—that while all the candidates for the Mathematical Tripos paid fees to the College for lectures in mathematics, whether they attended them or not, they did all they could to avoid going to these lectures, and relied on private tuition for their teaching. The system proposed by the College was that the College Lecturers, in addition to giving lectures, should take men individually, help them with their difficulties and advise them as to their reading. I was elected on the understanding that I should take part in this work. It was optional for those who were already Lecturers. The result was that I had most of the Trinity men coming to me, and as I kept on with some men from other Colleges who were already reading with me, I had about eighteen hours a week of mathematical teaching. I was only at this work for about two years. In one Tripos I had taught five out of the first six men in the list. After saying this, I ought to acknowledge that in one Tripos, the very last man, the wooden spoon, was a pupil of mine.[1] I think, however, that this was the greatest teaching triumph I ever had, for he was quite unable to follow any kind of mathematical reasoning. He could, however, learn pieces of book work off by heart, but without understanding them. I made him write out over and over again every piece of book work that was at all likely to be set in the elementary part of the Tripos until he could do them without mistake. This, however, did not make him safe, for there was no certainty that he could write out the right piece of book work in answer to a particular question, if the wording of the question differed to an appreciable extent from that to which he was accustomed. The issue became almost one of probability : if you have a number of balls, each with different numbers, and throw them at random into an equal number of holes, each hole having a number

[1] The "Wooden Spoon" was, as long as the Tripos list was arranged in order of merit, the man at the bottom of the list. He was so called because it was the practice for some of the members of his College to go to the Senate House, where he received his degree from the Vice-Chancellor, and immediately after he had done so to present him with a wooden spoon, or sometimes a wooden shovel emblazoned with the arms of his College and adorned with ribbons of the College colours.

corresponding to that on one of the balls, what is the chance that the number of balls which go into the right holes is not less than the number of questions you have to answer correctly to get through the examination ? Fortunately he had good luck and so obtained an Honours degree in mathematics in Cambridge University.

Though eighteen hours a week spent in teaching may be thought to leave little time for research, I am strongly of opinion that in general some teaching should be combined with research, and that the teacher should not regard his teaching as negligible in importance compared with his research. There is no better way of getting a good grasp of your subject, or one more likely to start more ideas for research, than teaching it or lecturing about it, especially if your hearers know very little about it, and it is all to the good if they are rather stupid. You have then to keep looking at your subject from different angles until you find the one which gives the simplest outline, and this may give you new views about it and lead to further investigations. I believe, too, that new ideas come more freely if the mind does not dwell too long on one subject without interruption, but when the thread of one's thoughts is broken from time to time. It is, I think, a general experience that new ideas about a subject generally come when one is not thinking about it at the time, though one must have thought about it a good deal before. It is remarkable that when ideas come in this way they carry conviction with them, and depose without a struggle ideas which previously had seemed not unsatisfactory. The psychology of the incidence of ideas must be a very interesting subject.

About the time I took my degree there were several

burning questions before the University which led to a great many heated and in some cases bitter discussions. The first of these was compulsory Greek. At that time and until many years afterwards no one could take a degree unless he had passed an examination in Greek. This regulation was not an old one, for until 1822 Greek was not obligatory. By 1880 the number of undergraduates who had done no Greek before coming to the University had increased so greatly that the regulation had become very irksome, and did not result in these students acquiring any appreciable knowledge of either the language or the literature of Greece. They crammed it up for a few months before the examination from translations of the set books, which followed as nearly as possible the order of the words in the Greek text, and which had no literary value. This modicum of Greek was forgotten in less time than it had been learned. A Grace was submitted to the Senate in November 1880, proposing that Greek be no longer compulsory, but was rejected by 180 votes against 140. In January 1919 a similar Grace was proposed and passed without any opposition.

A subject which excited even more interest, and which the undergraduates seized upon as an opportunity for indulging in very elaborate "rags", was the proposal to admit students from Newnham and Girton to University Examinations, that they should have the right to sit for the various Honours Examinations, and the place they took in the examination should be published in the list issued by the Examiners. This would not, however, entitle them to a degree. As a matter of fact there had already been cases in which, by the courtesy of the Examiners, women had been allowed to take the papers and these had been marked. Their place in the Tripos was not published in the University list, but it was not kept secret. In my year Miss Scott, who afterwards became Professor of Mathematics in Bryn Mawr University, U.S.A., was examined in this way in the Mathematical Tripos and took a high place among the Wranglers. The place she had taken had leaked out, and when the official list, in which of course her name did not occur, was read out in the Senate House, there was such an outburst of cheering when that place was reached, that it drowned the name of the man in the official list. He was in the Senate House, and as name after name was read out and his did not appear, he was in despair, and thought he must have made a dreadful mess of the examination.

There was a long and very animated discussion on the proposal in the Senate House, and for some days before the voting fly-leaves for and against the proposal were as thick as leaves in autumn. There was vigorous whipping-up on both sides and the non-residents came up in large numbers. The undergraduates took it up with great gusto : they paraded the streets, carrying banners inscribed with various sentiments, all, or nearly all, against the proposal. Figures supposed to represent women undergraduates in cap and gown were suspended across the streets. The proposal to allow women to take the Tripos papers and have their names published in the official list was carried by 398 to 366. The activities of the undergraduates lasted long after the declaration of the voting, for on the night of the voting I went with a friend after dining in Hall to smoke in Dr. Jackson's rooms, which were next to those of Mr Aldis Wright, the Vice-Master. We found a large goat trying to get past the Vice-Master, and forcing its way into his rooms. He had seized the goat by its horns, and the goat put down its head and butted, and the Vice-Master pushed. It was one of the most comical scenes I ever saw, for the Vice-Master was a very dignified old man, careful about his dress and with very stately manners. He was, too, in cap and gown, and his contest with the goat was about as incongruous as the Archbishop of Canterbury in full canonicals trying to drive an unwilling pig along a road. We went to his assistance, but he was evidently annoyed at being found in such a position and said he could manage by himself. All we could do was to tell a College porter to go and capture the goat, hoping that the Vice-Master would regard the removal of stray animals from the College as part of the porter's duties, and accept his assistance. Some of the raggers had smuggled a goat into College, driven it up the Vice-Master's staircase and then decamped.

The victory for the women was due to the admirable way in which Newnham and Girton had been managed, and to the popularity of their principals, Miss Clough and Miss Bernard (afterwards Mrs Latham). When Newnham opened in 1871 in a small house in Panton Street, the earlier students were the pioneers of University education for women, and they had among them, as all pioneering bodies have, some who went their own way without troubling much about what people said about them. They dressed in the way that seemed to them best and did not follow the fashions. This made them conspicuous, exposed them to a good deal of ridicule, and hampered the progress of the movement for University education for women. To alter this without giving offence must have been a very delicate business, but Miss Clough was not only very charming and capable but also very tactful. She is reported to have said to a student who looked as if she never brushed her hair, " My dear, you have very pretty hair, but I should like to see if it would not look still prettier if you did it in a way I will show you ". Whatever may have been the way she did it, she gradually reduced these eccentricities of dress until, at any rate to male eyes, they almost vanished. The popularity of Miss Clough and Miss Bernard induced some members of the Senate who did not believe in co-education in the abstract to say, if these ladies can turn out pupils like themselves it will be a very good thing, and we will give them a chance of doing it. No further steps were taken with regard to the position of women in the University until about the middle of the nineties, when a Syndicate was appointed to consider this question. They reported in favour of granting the titular degree of B.A. to those women who had fulfilled certain conditions of residence, had passed the examination such as the Little-Go which men have to pass before they can take a Tripos, and had obtained honours in some Tripos. The titular degree entitled the holder to put B.A. after her name, but did not carry with it the other rights and privileges which the B.A. degree gave to men. The argument put forward in favour of it was that, as is no doubt true, the general public who have not taken a degree themselves attach much more importance to B.A. after a person's name than those who have. Head mistresses of schools therefore prefer a belettered staff, so that students of Newnham and Girton, since they could not put B.A. after their names, were at a disadvantage in obtaining appointments. Even if they had not to seek for employment it was a pardonable vanity which made them wish for some simple way of showing that they had passed a Tripos. The proposal, however, met with the most determined opposition. The debate on it in the Senate House lasted for three days, and the report of the speeches occupied sixty-six pages of the *Reporter*. There was an avalanche of fly-sheets. The editor of the *Cambridge Review* took a postcard vote among undergraduates and resident B.A.s as to whether women should be admitted to the University (this, however, was not proposed by the Syndicate) and he sent out 2803 postcards and got 2169 replies, of which 466 were favourable and 1723 hostile. On the day of the voting there was a great influx of non-residents and the proposal was thrown out by the crushing majority of 1707 to 661. I believe the number of voters has never been equalled. The undergraduates thoroughly enjoyed the fight : they could not vote, but everything they could do to defeat the proposal they did, and celebrated the result with a stupendous " rag ". There is still (1936) in the windows of a well-known shop in the Market Place a photograph inscribed " The historic rag of 1897 ".

No further steps were taken on this question until 1920, when a Syndicate of twelve members was appointed to report upon it. The members had been chosen with such care to give each side an equal representation that they had to issue two reports, A and B, each with six signatures. Report A was in favour of admitting women to the University on the same terms as men ; they were to have the same degree (not a titular one but one carrying all the rights and privileges possessed by men). Report B was against this and thought the solution was to create a university or universities confined to women, like Bryn-Mawr, Smith, Vassar and Wellesley in America, which had been remarkably successful. Both these schemes, like the earlier ones, aroused fierce opposition. There were again long debates in the Senate House and sheaves of fly-sheets. The question was discussed by the undergraduates at a debate in the Union on the motion " This House does not consider that the granting of titular degrees without membership of the University meets the legitimate aspiration of women students ". This motion was lost, 185 voting for it and 375 against. There was again a very large vote in the Senate House and the proposal to admit women to membership of the University was rejected, 712 voting for it and 904 against. The behaviour of some of the undergraduates after the poll was declared in the Senate House was exceptionally deplorable and disgraceful. A large band of them left the Senate House, proceeded to Newnham and damaged the bronze gates which had been put up as a memorial to Miss Clough, the first Principal. There were rumours that this was at the instigation of an M.A. who, when the poll was declared, shouted from the steps of the Senate House, " Now go and tell Newnham and Girton ". This outrage aroused great indignation in the University among the undergraduates as well as the older members. The undergraduates at every College organised meetings, and at each of these a resolution was passed condemning " such contemptible behaviour and in particular that of a small number of undergraduates which culminated in disgraceful episodes and wanton damage". The authorities at Newnham dealt with this matter in a very dignified and charitable way.

These rags are nearly always very stupid and not infrequently lead to " regrettable incidents ". The psychology of a rag is that of a crowd, and this is not very different from that of a lunatic asylum. Some of the raggers really suffer from mild, temporary insanity. It is surprising how much some of them are affected by excitement alone without any aid from wine. I once at a bump supper, or something of that kind, sat next an undergraduate who was a bigoted teetotaller and to my knowledge drank nothing but ginger beer, and yet before the proceedings ended he was behaving as if he were very drunk indeed. It is the talking and shouting which brings this about, and a crowd can actually get intoxicated with the exuberance of its own verbosity alone.

With regard to the women's question. There was a widespread feeling among those who had voted against the proposal that something should be done to remedy the disability from which the women suffered in not having a symbol to indicate that they had taken a degree. Some, too, were influenced by the fact that there was a Royal Commission sitting on the University, and it was possible that if the University did not settle the question for itself the Commission would do it for them.

A petition with many signatures was sent to the Council of the University asking them to take steps to settle the matter. The Council, after much consideration, issued proposals of their own which granted a titular degree to women, and allowed them to be members of Syndicates and Boards and to hold educational posts in the University such as Professorships or Lectureships. The grant of a titular degree had been rejected by a great majority in 1897, but this time it was accompanied by the vitally important condition that the number of women students at Newnham and Girton should not, except for a few oddments, be increased beyond five hundred without a special Grace of the Senate. This means that unless the University itself determines otherwise, the proportion of women to men will be roughly one to ten, unless there is a considerable change in the number of male undergraduates.

In the debate in the Senate House the scheme was criticised by the extremists on both sides, but they did not carry their opposition to the length of voting against it, and when it was brought before the Senate it was carried *nemine contradicente*. The scheme has now been in operation for fourteen years and on the whole has worked well. Women have been appointed as Examiners in various University examinations ; they have been appointed Lecturers in several faculties, and I have never heard anything but praise of the way they have done their work. Though the wave of feminism has almost died away, I do not think there is any desire on the part of those who do not believe in co-education to break away from the compromise of 1921. One curious anomaly is that though women have no vote in the domestic affairs of the University, they can vote in the election for representatives of the University in Parliament.

Until the early eighties, College Fellowships had to be vacated on marriage if the Fellow were not a Professor in the University, or Registrary, Librarian or Public Orator. In 1882 the Colleges altered this and made marriage no longer a bar to holding a Fellowship. This had a profound effect on social life in Cambridge. Until then the only University families resident in Cambridge were those of the Heads of Houses, Professors, the Registrary and a few clergymen, doctors or lawyers who happened to be members of the University. I doubt if there were more than sixty such families all told. It was then a very small society and it was distinctly an elderly one, as the great majority had married late in life, so that it contained very few young men or women. My wife, who was born in Cambridge, and who has lived there all her life, says that there were certainly not more than ten young women among them in her girlhood. As soon, however, as the bar against marriage was removed there was a stampede to matrimony. Many of the younger Fellows were already engaged, and were only waiting to get some appointment away from Cambridge to get married. When they could get married without losing their Fellowships and College appointments they did so, and there was a great influx of young brides into Cambridge. This not only greatly increased the number of University families resident in Cambridge, but changed the society from one which was decidedly elderly to one exceptionally youthful. The chief form of entertainment before the change had been elaborate and rather formal dinner parties ; with the coming of the brides lighter forms of entertainment came into vogue, dances, lunches, teas, river picnics and so on.

The great increase in the number of Professors, Readers and Lecturers which has occurred since the war has also greatly increased the number of University residents. In 1882 there were 27 Professors, 2 Readers, 7 Demonstrators and 5 Teachers. In 1934 there were 72 Professors, 29 Readers, 229 University Lecturers and 35 Demonstrators. In addition, many more laboratories requiring superintendents, museums requiring curators, and libraries requiring librarians, have been founded. I think that the number of heads of families engaged in University work must have increased tenfold since 1882, and from being a very small society it has grown into a very large one.

Researches, 1880–84

After my election to a Fellowship, I began some mathematical investigations on moving charges of electricity. I was attracted to this by the beautiful experiments on cathode rays which had lately been made by Crookes, and

I was anxious to see what the behaviour of moving particles ought to be on Maxwell's theory that magnetic forces could be produced not only by electric currents through wires, but also by changes in electric force in a dielectric. Charges of electricity did not figure at all prominently in this theory, and when Maxwell in discussing electrolysis had for convenience of description to speak of a molecule of electricity, he says the phrase is out of harmony with the rest of his treatise. Helmholtz, who was a supporter of the theory, said he should be puzzled to explain what an electric charge was on Maxwell's theory beyond being the recipient of a symbol.

Since the electric force near an electrified particle varies very rapidly as the particle moves about, the behaviour of a moving charged particle would seem to afford a satisfactory method of testing the theory. I worked out, therefore, what on this theory would be the magnetic force produced by the moving particle, and what would be the mechanical force on the particle if it were acted upon by an external magnetic field. The results were published in the *Philosophical Magazine* for April 1891. I shall have to refer to them again when I am reviewing the progress of physics later on. The calculations in the paper were long, but there is one result which can be obtained without calculation which has formed the basis of a great deal of my work. The magnetic force is on Maxwell's theory proportional to the rate of change in the electric force. If e is the charge and v the velocity, the electric force at a given point will be proportional to e, and its rate of change to ev; hence at any point there will be a magnetic force proportional to ev; but where there is magnetic force there is energy, and the amount of the energy per unit volume is proportional to the square of the magnetic force. Hence the energy in the space round the moving charge will be equal to a quantity Ae^2v^2, where A is a positive quantity depending on the shape and size of the charged body. If the particle were not charged its energy would be $mv^2/2$, m being the mass of the particle. Hence the total kinetic energy of the charged body is $\left(\dfrac{m}{2} + Ae^2\right)v^2$, that is, its kinetic energy and therefore its behaviour under forces is the same as if its mass were not m but $m+2Ae^2$. Hence the mass has been increased by the charge, and since the increase is due to magnetic force in the space around the charge, the increased mass is in this space and not in the charged particle. It is interesting to compare this result with that for a sphere moving through water. When the sphere moves it sets the water around it in motion. The necessity of doing this makes the sphere behave as if its mass were increased by a mass equal to half the mass of a sphere of water of the same volume as the sphere itself. This additional mass is not in the sphere but in the space around it. Now no one supposes that the dynamics of the sphere moving through water is not Newtonian dynamics, and there is no reason to suppose that the Newtonian dynamics can not apply to the motion of the charged particle ; it is not the dynamics which need be altered, it is the place where the mass must be located. We must suppose that the place where the mass must be found is in the space surrounding the charged particle, and not in the particle itself. If we adopt the electrical theory of the constitution of matter we may suppose that all mass is electrical in its origin, and therefore not in the atoms or molecules themselves but in the space around their charges. The hydrodynamical analogue of this would be the case of the motion through water of exceedingly thin spherical shells, so thin that their mass was infinitesimal in comparison with that of the water they displaced. A collection of these would have a finite mass equal to half that of the water they displaced.

The subject for the Adams Prize for 1882, a prize established to commemorate the discovery of Neptune by Professor Adams, was announced about the time I had finished the paper on the moving particle. It was an investigation on " the action of two vortex rings on each other ". I was greatly interested in vortex motion since Sir William Thomson had suggested that matter might be made up of vortex rings in a perfect fluid, a theory more fundamental and definite than any that had been advanced before. There was a spartan simplicity about it. The material of the universe was an incompressible perfect fluid and all the properties of matter were due to the motion of this fluid. The equations which determined this motion were known from the laws of hydrodynamics, so that if the theory were true the solution of the problem of the universe would be reduced to the solution of certain differential equations, and would be entirely a matter of developing mathematical methods powerful enough to deal with what would no doubt be very complex distributions of vortex motion in the fluid. It seemed well worth trying if there were any cases we could solve, and finding whether these involved anything inconsistent with the properties of matter. Another thing that appealed to me was the analogy between the properties of vortex filaments and those of the lines of electric force introduced by Faraday to represent the electric field. Faraday's lines, like vortex filaments, could not be created nor destroyed, and they must also end on electric charges ; a vortex filament must end on the boundary of a fluid and we might conceive the electric charge acting as a boundary. In fact it seemed that even if the vorticity did not suffice to represent matter, it might yet give a very useful representation of the electric field. The investigation of the problem set for the Adams Prize, like most problems in vortex motion, involved long and complicated mathematical analysis and took a long time. It yielded, however, some interesting results and ideas which I afterwards found valuable in connection with the theory of the structure of the atom, and also of that of the electric field. The essay was awarded the prize and was published in 1883 by Macmillan under the title *A Treatise on the Motion of Vortex Rings.* In the same year I gave, in a paper published in the *Proceedings* of the London Mathematical Society, the solution of the problem of finding the electrical oscillations which can occur on the surface of a conducting sphere.

I began to work in the Cavendish Laboratory immediately after taking my degree. I had not done so before, though I used to go to the laboratory now and then to see Poynting or Schuster, who were working there. I was not fortunate enough to meet Maxwell on any of these occasions : he was then editing the unpublished papers of Henry Cavendish, and not engaged on any regular research. Maxwell's view of the function of the laboratory was that it should be a place to which men who had taken the Mathematical Tripos could come, and, after a short training in making accurate measurements, begin a piece of original research. Most of the men who were working there when I first knew it were of this type. There was no organised teaching of undergraduates, though some lectures for candidates for the Natural Science Tripos and for medical students were given by the Demonstrator, William Garnett, afterwards Educational Adviser to the

London County Council. He had been a candidate in the Mathematical Tripos when Maxwell was an Examiner, and had so impressed him by the soundness of his work on the physical questions that he made him his Demonstrator. No fees were charged for working in the laboratory. Maxwell's own lectures in the laboratory were not well attended. Indeed, in the last year the class consisted of an American and the eminent physicist who is now Sir Ambrose Fleming. I never heard Maxwell lecture in the laboratory, but I heard him in 1879 give the Rede Lecture in the Senate House on the Telephone. It was in the very early days of the telephone. The working of the telephone was demonstrated by an experiment in which a tune was played in the Geological Museum and heard in the Senate House : the distance between the buildings was less than twenty yards. It was a very pleasant lecture to listen to : there were many instances of that " pawky " humour which flashes out so often in his verses. The Vice-Chancellor was not very happy in proposing the vote of thanks. He said they were much indebted to Professor Maxwell, for he had helped them when they were in a great difficulty. They had asked everyone they could think of to be Rede Lecturer and they had all refused, and he did not know what would have happened if someone at the last moment had not suggested Professor Maxwell.

Maxwell died in 1879 and was succeeded by Lord Rayleigh, who began work in the Lent Term of 1880. The work of the laboratory was much extended; a scheme for instruction to undergraduates both in theoretical and practical work was organised. Glazebrook and Shaw were appointed Demonstrators, and to meet the cost of the apparatus required for these developments, Lord Rayleigh raised a sum of £1500. All the new developments prospered, and the number of undergraduates studying in the laboratory increased rapidly.

I began to work in the laboratory, in 1880, by attempting to detect the existence of some effects which I thought would follow from Maxwell's theory that changes in electric forces in a dielectric produced magnetic forces. The results I obtained were not sufficiently definite to allow any positive conclusions to be drawn from them, and I then took up a research suggested by Lord Rayleigh. It was on some effects produced in the working of induction coils by the electrostatic capacity of the primary and secondary of the induction coil. An account of the research was published in the *Philosophical Magazine* for July 1881. I then, again at the instance of Lord Rayleigh, began a much more difficult and lengthy research on the determination of the ratio of the electrostatic to the electromagnetic units of electric charge which ought, on Maxwell's theory, to be equal to the velocity of light ; the work of previous observers showed that the difference was not great. About this time I was a candidate for the Chair of Applied Mathematics which had just been established at Owens College, but was not successful. This was not surprising as the successful candidate was Arthur Schuster, who, like myself, had been a student at the College, was already teaching there and doing it very successfully. In 1883 I was elected a University Lecturer and in that capacity lectured on electrostatics and electromagnetism in the academic year 1883-84. I also in the same year gave two courses of lectures at Trinity College, one on rigid dynamics and the other on statics and attractions, which were open to all members of the University. In the spring of 1884 I was elected a Fellow of the Royal Society. In the summer of

1884 Lord Rayleigh, who had stipulated when he accepted the Cavendish Professorship in 1879 that he must not be expected to hold it for more than five years, resigned the Professorship, and in December 1884 I was, to my great surprise and I think to that of everyone else, chosen as his successor. I remember hearing at the time that a well-known College tutor had expressed the opinion that things had come to a pretty pass in the University when mere boys were made Professors. I had sent in my name as a candidate without dreaming that I should be elected, and without serious consideration of the work and responsibility involved. When after my election I went into these, I was dismayed. I felt like a fisherman who with light tackle had casually cast a line in an unlikely spot and hooked a fish much too heavy for him to land. I felt the difficulty of following a man of Lord Rayleigh's eminence. I remembered that I had never given lectures at which experiments had to be performed, and that I had never taught any classes in practical physics. Happily my want of experience in this respect was made less harmful than it would otherwise have been by the kindness of Glazebrook and Shaw, who continued to take charge of the classes in practical physics which had been organised by them when Lord Rayleigh was Professor.

CHAPTER IV

The Cavendish Laboratory
and Professorship of Experimental Physics

THE Cavendish Laboratory has played such a large part in my life that I hope to be pardoned if I say something about its history. Though many very great discoveries in physics had been made in Cambridge as far back as Newton's time, it was not until 1869 that steps were taken to have in the University a physical laboratory and a Professor of Experimental Physics. Newton and Stokes had made their experiments in their own rooms with their own apparatus and at their own cost. In the early sixties, however, the importance of science in education was warmly advocated by Huxley, Roscoe and Lockyer along with others, and considerable interest in the teaching of science was aroused. In 1866 Clifton was appointed to a new Chair of Experimental Physics in Oxford and the building of the Clarendon Laboratory was commenced in 1868 and finished in 1872. In Cambridge there was in addition to the general recognition of the importance of science in education a special one for making provision for the teaching of heat, light, electricity and magnetism, for in 1868 these subjects had been added to those included in the Mathematical Tripos, and it was necessary to provide for instruction in them. A Syndicate appointed in November 1868 to consider how this might best be done, reported in February 1869 in favour of establishing a Professorship and Demonstratorship of Experimental Physics, and the erection of a physical laboratory adequately supplied with apparatus ; they thought this would cost £6300. The University was then so poor that it was clear that there would be great difficulty in raising this sum. The situation was saved, however, by the Chancellor of the University, the 7th Duke of Devonshire—who had been first Smith's Prizeman and Second Wrangler—offering to provide funds for building the laboratory and stocking it with apparatus. The building alone cost £8450, considerably more than the estimates, and the Duke not only defrayed this, but also continued to provide apparatus for the laboratory until Maxwell, three years after the laboratory had been opened, reported " that it contained all the instruments required by the present state of science ". It has never since been in this condition.[1] If it had not been for the generosity of the Chancellor, though no doubt in time the University would have had a physical laboratory, it would not have had Maxwell as a Professor, and the Cavendish Laboratory would have been without the inspiration and tradition which it owes to its first Professor. In February 1871 the University sanctioned the creation of a Professorship in Experimental Physics, to which Maxwell was appointed in March of the same year. It was believed at the time that the University had first approached Sir William Thomson and then von Helmholtz, the great German physicist and physiologist, but neither of these could see his way to accept the post. At the time of his election Maxwell's work was known to very few, and his reputation not comparable with what it is now. The *Treatise on Electricity and Magnetism* did not appear until two years later, and

[1] The list of these instruments was published in the *Cambridge University Reporter* (May 15, 1877). It is a striking instance of the difference between the apparatus which was then considered adequate, with what would be so now.

though he had published the fundamental ideas long before
in the scientific Journals, they had attracted but little attention, and his reputation was based mainly on his work on
the kinetic theory of gases. Indeed, even at the time of
his death the truth of his supreme contribution to physics
—the theory of the electromagnetic field—was an open
question. It was only when nearly ten years later Hertz
detected by experiment the electromagnetic waves, which
were the characteristic and essential part of his theory, and
which distinguished it from all others, that the importance
of his work was adequately realised. Maxwell delivered
his inaugural lecture in October 1871. It is a very able
and interesting essay on the functions of experimental
work in the laboratory in University education. It has
been quoted very frequently, perhaps oftener than any
other of his writings. But for the moment it was almost a
fiasco. Sir Horace Lamb, who was at it, describes it in the
James Clerk Maxwell Commemoration Volume : " The
announcement of it had been made in such a way that it
had escaped the notice of the older members of the University. It was not given in the Senate House, the usual place
for such lectures, but in an obscure lecture-room. The
result was that only about twenty were present, and these
were all young mathematicians who had just taken, or were
about to take, the mathematical Tripos, but more remains
behind." I quote Sir Horace Lamb's account of this :
" The sequel was rather amusing. When, a few days later,
it had been announced with proper formality that Professor Maxwell would begin his lectures on heat at a certain
time and place, the *dii majores* of the University, thinking
that this was his first public appearance, attended in full
force out of compliment to the new Professor, and it was
amusing to see the great mathematicians and philosophers

of the place such as Adams, Cayley, Stokes, seated in the front row while Maxwell, with a perceptible twinkle in his eye, explained to them the difference between the Fahrenheit and Centigrade scales of temperature."

"It was rumoured afterwards, and perhaps it is not incredible, that Maxwell was not altogether innocent in this matter, and that his personal modesty, together with a certain propensity to mischief, had suggested this way of avoiding a more formal introduction to his Cambridge career."

During the erection of the laboratory he lectured each term in such rooms as happened to be vacant. He said he went about like a cuckoo, dropping ideas about heat in the chemical laboratory in the October Term, about electricity in the botanical laboratory in the Lent Term, and about magnetism in the New Museums in the Easter Term. His lectures were quite elementary and were not so well attended as they ought to have been. Some who attended have expressed the pleasure they got from the quips and cranks and dry pawky humour which now and then broke through the crust of science. This side of his character comes out very clearly in the verses which he wrote from time to time, and which are published at the end of Lewis Campbell's biography.

The laboratory was formally opened by the Chancellor of the University in October 1874. I believe W. M. Hicks, who was afterwards Professor and then Principal at Sheffield University, was the first to come, but others gradually drifted in, most, though not all, men who had lately taken the Mathematical Tripos and had no previous experience of experimental work. At first they were set to read scales and verniers, to measure times of vibrations, to use a reflecting galvanometer, to measure the resistance

Photo : Turner & Sons, Cambridge

THE CAVENDISH LABORATORY

The building on the extreme left is the new extension.

of a wire ; after a short time spent in this way, they were often set to measure the horizontal component of the earth's magnetic force by a magnetometer of the Kew pattern. Maxwell had a high opinion of the training to be got by using this instrument. It afforded practice not only in reading scales and measuring times but also in making adjustments, and the accuracy with which these had been done was indicated by the value obtained for the magnetic force. After a short course of this kind they began to work at a specific problem. He liked the student to have thought of one for himself. " I never," he said, " try to dissuade a man from trying an experiment. If he does not find out what he is looking for he may find something else." In most cases, however, the student wished Maxwell to select a subject for him. Maxwell did so, and spent great pains in planning how the experiment had best be tried. When he was satisfied with this, he acted on what I believe to be the right principle of encouraging the student to try to overcome the difficulties himself, and that it is better for the teacher to do this rather than to remove them out of his way. Maxwell went round the laboratory for an hour or two most days, generally accompanied by his dog, and talked with the students about their researches. They also got valuable help from the Demonstrator, W. Garnett, who was a good amateur carpenter, and was always ready to help those who were not, in making the alterations in the apparatus or the simple contrivances which are continually being called for in experiments. This help was especially valuable, for at the beginning there was no skilled workman in the laboratory. At first all the students were research workers, nearly all of them graduates. The first undergraduate who worked in the laboratory was my old friend H. F. Newall, who later became Professor of Astro-Physics at Cambridge. He in his first term, against the advice of his tutor, entered this unexplored region, and like most explorers found the going was not easy.

Important researches were made between 1874 and 1879 in the laboratory by G. Christal, later Professor of Mathematics in Edinburgh University, on the accuracy of Ohm's Law; by Donald MacAlister, Senior Wrangler and later Sir Donald MacAlister, Principal of Glasgow University, on a new proof that the attraction of a point charge of electricity varies inversely as the square of the distance; by R. T. Glazebrook (Sir R. T. Glazebrook, K.C.B., F.R.S.), later Principal of Liverpool University, after that Director of the National Physical Laboratory, on the wave surface in a biaxial crystal; by Arthur Schuster, afterwards Sir Arthur Schuster, F.R.S., in spectroscopy; by J. A. Fleming, now Sir Ambrose Fleming, on the measurement of resistances. Maxwell himself did not do any continuous experimental research in Cambridge. Though he came to Cambridge in 1871, the laboratory was not opened until 1874. Afterwards, the greater part of his time was spent in editing the works of Henry Cavendish, a pious duty for a Cavendish professor. Cavendish, though he only published two papers, left twenty packages of manuscripts on mathematical and experimental electricity. Maxwell copied these out with his own hand. He saturated his mind with the scientific literature of Cavendish's period; he repeated his experiments; he was especially attracted by one where Cavendish had been his own galvanometer and had estimated the strength of the current by the shock it gave him when he passed it through his body. Visitors to the laboratory had currents passed through them to see whether or not they were good galvanometers. The papers were published in 1879 under the title, *The Electrical Researches of the Honourable Henry Cavendish*. They showed that Cavendish was even a greater man than had been thought. He anticipated Faraday in the discovery of the property known now as Special Inductive Capacity and measured its values. He had formed the conception of electrostatic capacity, had anticipated Ohm's Law, had made experiments which proved that the force exerted by a point charge of electricity varies inversely as the square of the distance from the point. These papers confirm the impression produced by those published in his lifetime, that he was a superb experimenter. He had the gift, like Lord Rayleigh, of seeing what was the vital point in the experiment, and though the apparatus as a whole might look untidy and haphazard, the parts that really mattered were all right. It was a " rum 'un to look at but a beggar to go ". A striking feature in these papers of Cavendish is that they are quantitative; definite measurements are given of the various effects. A great deal of the work done by the earlier experimenters had been, as was natural in the infancy of the science, qualitative. They tried to find whether an effect got greater or less under a certain change of conditions, but not how much greater or how much less. Cavendish, who measured everything he had to do with, measured electricity and speaks of inches of electricity. His researches on electricity were made before the death of his father, Lord Charles Cavendish, who was also interested in science, in 1783. Until then he had lived with his father, and his apartments were a set of stables fitted up for his accommodation. After his father's death he moved to a house in Clapham, the greater part of which was used for his experiments. It was here that he made his discovery of the composition of water, and measured by means of the torsion balance the density of the earth. There is no evidence that he made any more electrical experiments. It was natural that Cavendish in his earlier researches should have worked at electrical subjects, for at that time they were the ones which attracted far more attention than any others. The discovery of the Leyden jar, Franklin's researches, particularly the invention of the lightning conductors, had aroused such an interest in electricity, and provided so many subjects for research, that they seemed the most promising, as they were the most popular, ways of obtaining new and important results.

Cavendish's life was dominated by science to an extent which seems almost incredible. From the evidence of his contemporaries, given in Wilson's biography, he never talked about anything else. In spite of his family traditions, he never took any interest in public affairs. In the latter part of his life he was very wealthy, yet he never paid any attention to his financial matters. When his balance at the bank was £80,000, his banker thought he ought to go to him and advise him to invest some of it, instead of letting it lie idle, earning no interest. He went, and Cavendish was very angry and said, " What do you come here for ? " He told him. Cavendish said, " If it is any trouble to you I will take it out of your hands, so do not come here to plague me. What do you want me to do ? " " Perhaps you would like to have £40,000 invested." " Do so, and don't come here to trouble me or I will remove it."

He never wasted any time in deciding what he should do in the private affairs of life ; he always did what he had done before. He took the same walk at the same time every day of his life, walking in the middle of the road to avoid meeting anybody he might chance to pass on the footpath. His tailor provided him on a fixed day in the year with a new suit which was a replica of the old one. He was a misogynist as well as a recluse. His female servants were there on the understanding that they should keep out of his sight ; if he saw them they were dismissed. Even with his housekeeper his communications were written, not spoken.

The occasions when he went into society were all within the ambit of the Royal Society, to which he had been elected in 1760. He attended regularly its meetings, the receptions given on Sunday evenings by its President, Sir Joseph Banks, and above all the meetings of the Royal Society Club, a dining-club connected with the Royal Society which is still flourishing. It appears from the information contained in *The Annals of the Royal Society Club* by Sir Archibald Geikie, that at the meetings of this club, when he was surrounded by men of science, he was a very different person from what he was in ordinary life. There is in existence an account of a club dinner at which Cavendish was present, written by a French guest. After describing the dinner itself, when nothing but porter was drunk, " and this out of cylindrical pewter pots, which are much preferred to glasses because one can swallow a whole pint at a draught ", he goes on to speak of the number of toasts which were drunk after the cloth had been removed, and the wine was going round. After these had been drunk and followed by brandy and rum, he says, " I repaired to the Society [the Club dinners began at 5, and the meeting of the Society at 8] along with Messrs Banks, Cavendish, Maskelyn, Aubert and Sir

Henry Englefield. We were all pretty much enlivened but our gaiety was decorous."

Cavendish was made a member of the Club in 1760, a few months after he had been made a Fellow of the Royal Society, and in accordance with his habit of doing the same thing over and over again, he attended with a regularity which has never been even approached by anyone else. In those days the Club met every Thursday throughout the year, much more frequently than it does now.

From 1770 onwards until the end of his life, his record was never lower than 44 attendances in the year, and was usually about 50. In 1784 January began on a Thursday and December ended on Friday, thus allowing 53 dinners in the year, and he was present at every one. He always put his hat on the same peg.

He not only came himself, but also frequently brought guests. One year he had as many as nineteen. These were not all physicists or mathematicians ; some were doctors, some engineers, travellers, naval officers. Professor Playfair, who was a guest at these dinners, says, "He never speaks at all but that it is exceedingly to the purpose, and either brings some excellent information or draws some important conclusion. His knowledge is very extensive and very accurate. Members of the Royal Society look up to him as one confessedly superior."

Dr. Wollaston, who also often dined at the Club, said, " The way to talk to Cavendish is never to look at him but to talk as it were into vacancy, and then it is not unlikely you will set him going ". His silence in general society may have been because he had nothing to say about the things which interested his audience. I have known more than one man of science who would sit through a long dinner without saying a word to those sitting next to him, and yet would speak freely and well on things that interested him.

In the case of Cavendish we must remember that the motto of the Royal Society is *Nullus in verba,* and his family motto *Cavendo tutus,* and neither of these encourages loquacity.

Maxwell's health began to fail at the beginning of 1879 and he was able to spend only very little time in the laboratory. He went to Glenlair, his house in Galloway, for the summer but rapidly grew worse. He was brought back to Cambridge to be under the care of his favourite physician, Sir George Paget. He suffered much pain, which he bore with great fortitude. He died on November 5, 1879.

The Professorship had been established under the regulation that it was to "terminate unless the University by the Grace of the Senate shall decide that the Professorship shall be continued". A Grace to continue the Professorship was passed on November 20, 1879. Lord Rayleigh was persuaded to become a candidate and was elected on December 12, 1879. He began his work at the laboratory in the Lent Term of 1880, giving a course of lectures on physical apparatus which were well attended. He also began to organise and develop the teaching of physics to undergraduates, which had hitherto been somewhat neglected. The need for this was pressing, for the number of students taking physics was increasing quickly, and also because the University had decided that in 1881 the Natural Sciences Tripos was to be divided into two parts. In Part I candidates were expected to take the more elementary parts of three or more subjects, while in Part II they were expected to specialise in one subject. This Part would demand a knowledge of the more advanced parts of physics, and for this there was as yet no provision. To provide it would require a good deal of extra apparatus, and one of the first things Lord Rayleigh did was to raise a sum of about £1500 for this purpose by asking for donations from a number of his friends. Garnett resigned his Demonstratorship in the Lent Term of 1880, and R. T. Glazebrook and W. N. Shaw were appointed Demonstrators. They, in consultation with Lord Rayleigh, evolved a scheme for instruction in practical physics which, in its essentials, was the same as that now in force more than fifty years later. There were demonstration classes each term on some branch of elementary physics; these were three times a week, each lasting for two hours. The subjects included in Part I of the Natural Sciences Tripos were covered in three terms. A student was required to finish one experiment and write out in his notebook an account which satisfied the Demonstrator before he could proceed to another. A similar plan was adopted for the demonstrations in the advanced subjects. The students were not limited here, however, strictly to two hours, but could stay longer if they wished. There was a laboratory notebook for each experiment, and the students wrote the account of their work in this as well as in their own book. This plan worked well; the student thought he was writing for posterity and took special pains with his description of his experiment. This was good practice in a very important thing—clear exposition of the results he had obtained. The numbers attending the elementary classes increased so rapidly that, before the end of Lord Rayleigh's tenure of the Professorship, they had to be duplicated, and two additional Demonstrators appointed to assist Glazebrook and Shaw.

Almost as soon as he came to Cambridge, Lord Rayleigh

began the determination of the absolute measure of various electrical quantities, which raised the standard of electrical measurement to a higher plane. He began with the re-determination of the ohm. There had been a determination of this made in 1864 under the direction of a Committee of the British Association. The experiments were made by Maxwell at King's College, London, where he was then Professor. They issued a standard, afterwards called the B.A. Ohm, which professed to have a resistance in absolute measure equal to 10^9. Later experiments had led to serious doubts about the accuracy of this standard, the differences covering a range of about 3 per cent. The apparatus used by the Committee was in the Cavendish Laboratory, and was used by Rayleigh in collaboration with Mrs Sidgwick and Schuster. The apparatus was set up, and experiments commenced, in the summer of 1880 and the experiments concluded by the end of 1881. The result was that the resistance in absolute measure of the B.A. unit was $\cdot987 \times 10^9$ instead of 10^9 as it ought to have been. The greater part of the error was traced to a very commonplace cause, a mistake in arithmetic. In the experiments several instruments had to be observed simultaneously, and visitors to the laboratory were sometimes called on to take a hand. I remember seeing Mr Arthur Balfour set down to read a galvanometer.

Rayleigh did not stop here. He repeated the experiment, using the same method but larger apparatus, then he made another determination using a different method. The three determinations gave very concordant results. These experiments had taken three years of very difficult and often harassing work ; for example, at one stage the readings taken one day did not agree with those on the next, and it took two months to get over this difficulty.

Lord Rayleigh, in collaboration with Mrs Sidgwick, went on to determine in absolute measure the other electrical standards: the ampere, the standard of electric currents, and the volt, the standard of electromotive forces.

Lord Rayleigh, by the experiments he made when he was Cavendish Professor, raised the standard of electrical measurement to such a high level that it may be claimed that here he has changed chaos into order. His results have stood the test of time; they have been confirmed by later measurements made with larger and more elaborate apparatus, and have been adopted by International Congresses as the basis for the definition of the Ohm, Ampere and Volt for legal purposes. The accuracy of the value of the standards is of vital importance both to the physicist and to the electrical engineer. The physicist in his mathematical investigations must use the absolute measures of any electrical quantity which may occur in them, and express the results he obtains in terms of them. Suppose, for example, that his calculations indicate the existence of a certain effect whose magnitude depends on the electrical resistance of a piece of the apparatus he is using to detect it. If he wishes to test this in the laboratory, he has to measure this resistance in ohms. If the absolute measure of the ohm is wrong, the results he obtains will not be in accordance with those predicted by the theory; he may conclude that his theory is wrong and must be abandoned, whereas the difference is due to the error in the ohm. When Lord Rayleigh began his experiments there was an uncertainty of about 3 per cent in the absolute value of the ohm. This might cause a discrepancy of this amount between the theoretical and observed results, far greater than could be accounted for by errors in carefully made experiments. Again, the electrical engineer, like the physicist, uses the absolute value of electrical quantities in his calculations, and if the ohm, ampere or volt are wrong, the actual efficiency of a dynamo or motor will be different from the estimate. This difference in a huge industry like electrical engineering, when expressed in terms of money, may involve very large sums.

Besides his experimental works, Lord Rayleigh published many theoretical papers of first-rate importance during his stay in Cambridge. It was his custom not to come to the laboratory in the morning if he was not lecturing, but work in his study at home. I think the one that gave him the greatest pleasure was that on the soaring of birds, in which he gave an explanation of the striking fact that some birds, *e.g.* gulls and albatrosses, can, without moving their wings, and in the absence of an upward wind, soar round and round at a constant height, and even in some cases get higher and higher. Rayleigh showed that this might be possible even where the wind was horizontal, if its velocity increased with the height above ground. The bird must go downwards when flying with the wind and upwards when flying against it. I remember his talking to me about this before it was published ; in fact he felt some hesitation about publishing it because he thought it was too pretty to be correct. It is now, I believe, universally accepted.

There were several researches going on in the laboratory besides those made by Lord Rayleigh himself. Glazebrook determined the ohm by a method different from either of those used by Lord Rayleigh, and completed a research on the reflection of light from the surface of a uniaxial crystal. Shaw began his meteorological investigation on the comparison of different forms of hygrometers. Sir George and Sir Horace Darwin endeavoured to find evidence of tides in the earth's crust due to the motion of the moon, but the conditions were not sufficiently free from other disturbances to allow this effect to be detected.

No mention of the Cavendish Laboratory at this period would be complete without mention of Mr George Gordon, who was in charge of the workshop. He had been a ship-wright in Liverpool and was very skilful and quick with the adze. His work was not pleasing to the eye but it did what it was expected to do, which was all that Lord Rayleigh demanded.

We have been fortunate enough to have had as teachers in practical work in physics at the Laboratory, in addition to those I have mentioned: W. C. Dampier Whetham, F.R.S. (now Sir William Dampier); Professor H. L. Callendar, F.R.S., an intellectual Admirable Crichton, who took first classes in both the Classical and Mathematical Triposes and was awarded a Fellowship at Trinity College for Physics; the Rev. T. C. Fitzpatrick, who later became President of Queens' College, Cambridge; S. Skinner (later President of the South Western Polytechnic, Chelsea); C. E. Ashford (later Head Master, Royal Naval College, Dartmouth, now Sir C. E. Ashford), and many others.

Mr G. F. C. Searle, F.R.S., who first began to demonstrate in 1891, has done more than anyone else for the teaching of practical physics at Cambridge. It was only in the year 1935 that, owing to the age limit, he gave it up after forty-four years' uninterrupted service, teaching all the time the largest class—that for students who are going in for Part I of the Natural Sciences Tripos. Nothing approaching such a long tenure as this of the office of Demonstrator has ever come to my knowledge. In general the zeal for demonstrating fades away after a few years : the work is hard and may be monotonous, but Searle is as keen now as he was forty-four years ago; indeed it was only the other day (1935) that he invented a new and very beautiful experiment for his course. He has a real enthusiasm for the subject; he is not content with merely teaching it. He has taken infinite pains and thought to improve the course, and replace old experiments by others which have a greater educational value. His experiments are never commonplace : they have always a freshness and elegance which makes one want to do them. After he has got the idea, he works away at the experiment until he has found the form which gives the most accurate results, and so is suitable for training the students in accuracy of measurement. This often involves a great deal of work. He then writes out an account of the theory, followed by a description of the way the experiment is made. He takes infinite pains to make his meaning clear, making one attempt after another until he is satisfied. He then makes a fair copy in his own hand, and this is placed in the laboratory for the use of his students. He must have made, I should think, many more than a hundred of these. Since for more than forty years practically every Cambridge man taking physics has been a pupil of Dr. Searle, the influence he has exerted on the teaching of practical work in physics must have been comparable with that exerted by Routh on the teaching of mathematics.

A very important person in a physical laboratory is the man in charge of the workshop. The smooth running of the laboratory depends upon him almost more than upon anyone else, and to be successful he must have many qualifications. Besides being a good workman he must be a man of strong character, for he has to maintain discipline among the younger assistants. He has to be business-like, for the buying of stores and materials will be in his hands, and he requires the tact of a diplomat to reconcile the claims on the workshop of those engaged in teaching, and those working at research. One of my first duties on becoming Professor was to appoint such a man to superintend the workshops, for Gordon, who had held this post, left to be private assistant to Lord Rayleigh. I was fortunate in finding in a young Scotsman, Mr D. S. Sinclair, a man who combined all these qualities. Though a good mechanic, he knew nothing about glass-blowing when he came to the Laboratory. I sent him to have a few lessons, and after two or three months he became a very expert and exceptionally quick worker. He did very good work for the Laboratory for three years, when he left to take a good appointment in an engineering firm in India. He was succeeded by Mr A. T. Bartlett, who after five years' service left to take up electrical engineering. He now holds an important position as head of the Research Department of the English General Electric Company. After him came Mr W. G. Pye, the son of Mr Pye who managed the workshop established by Mr Dew-Smith for making apparatus for use in the Physiological Laboratory, and which developed into the Cambridge Scientific Company. W. G. Pye rendered us very efficient service from 1892 to 1899, when he left to start in business in Cambridge as a scientific instrument maker. This business was very successful. It specialised in producing well-made and well-designed apparatus at a moderate price suitable for school laboratories or for elementary classes in the university. An offspring of this, Pye's Radio, which bears his name, is a still larger concern giving employment to many hundred workers. He was succeeded in 1899 by our present assistant, Mr F. Lincoln, who has rendered invaluable assistance in the work of the

Laboratory for thirty-six years. Mr Lincoln came to the Laboratory in 1892 when he was a very small boy, so small that he had to stand on a box to reach the bench. W. H. Hayles, who afterwards became chief lecture assistant, was appointed by me in the first term of my Professorship. He rendered good service to the Laboratory for more than fifty years, when he retired under the age limit. He was a skilful photographer, and very expert in making lantern slides. He had the valuable quality of realising to the full what an important part he played in the lecture, and spared neither time nor trouble to make the experiment succeed.

My first course of lectures was given in the Lent Term of 1885. The beginning was not very auspicious, for in my first lecture, though I got on quite well until about ten minutes before the time to stop, I then became very dizzy: I could not see my audience, and thought I was about to faint and had to dismiss the class. I recovered in a few minutes and have never suffered in this way again. I attribute it to having turned round to look at the blackboard at my back too often. I have sometimes felt this in a slight form when lecturing in a darkened room and showing many lantern slides, and often turning my eyes from the brightly lighted screen to the darkened room.

Immediately after my election to the Professorship, I began, in collaboration with my old friend, Richard Threlfall, some experiments on the passage of electricity through gases. Threlfall had just taken a First Class in physics and chemistry in Part II of the Natural Sciences Tripos, but his enthusiasm for physics began long before he came to Cambridge. He had as a boy a laboratory and workshop at his home near Preston, and experimented during the vacations. When working at high explosives he had an explosion which blew off large parts of several fingers. In spite of this he was one of the best experimenters I ever met, and very skilful in the use of tools. I am myself very clumsy with my fingers, so that his assistance was of vital importance. This was the first of my experiments on the passage of electricity through gases, and since then there has, I think, never been a time in which I have not had some experiments in hand on that subject. I was attracted to it because I wished to test the view that the passage of electricity through gases might be analogous to that through liquids, where the electricity is carried by charged particles called ions. My view was that whenever a gas conducts electricity, some of its molecules must have been split up by the electric forces acting on the gas, and that it is these which carry the electricity through the gas. On this view, a gas in which all the molecules are in the normal state cannot conduct electricity. My idea at that time was that some of the molecules were split up into two atoms, one of which was positively, the other negatively, electrified, and my first experiments were intended to test this idea. It was not until 1897 that I discovered that the decomposition of the molecule was of quite a different type from ordinary atomic dissociation ; then I found that one of the bodies into which the molecule split up, the one carrying the negative electricity, is something totally different from an atom, and that its mass is less than one-thousandth part of that of an atom of hydrogen, the lightest atom known.

Towards the end of my first years of office, T. C. McConnel, Fellow of Clare College, a man with a singularly acute and original mind, and who had done valuable original work in optics whilst at the Laboratory, left Cambridge to take part in a large business in Manchester with which his family were connected. Unfortunately his health broke down shortly afterwards and he spent the rest of his life in Switzerland, where he made valuable observations on the properties of ice. He explained the bending of ice, in cases when regelation was impossible, by the slipping of one crystal over another. He was one of the first to realise the importance of slip, and the difference between the properties of a single crystal and an aggregate.

Threlfall succeeded McConnel as Assistant Demonstrator, but was shortly afterwards appointed Professor of Physics in the University of Sydney, N.S.W. His first experiences were very characteristic ; as soon as he arrived in Sydney he drove to the University to see the laboratory, but when he arrived there he was told there was no laboratory. Then he said he was going back to England, as he was not going to be a Professor of Experimental Physics without a laboratory where he could make experiments. They took him over the building, and showed him a room here and another there which he might use for his laboratory, but he said this would not do ; he must have a proper laboratory. Then they promised to apply to the Government and use all their influence to get it to build a laboratory for the University. Then he agreed to stay, and broke to them that, as he knew he would not be able to get physical apparatus in Sydney, he had ordered in London apparatus which would cost about £2000, and which would be coming over before long. For some months the Government did nothing towards providing a laboratory, but in the meantime Threlfall had become very friendly with the Prime Minister, who was very fond of working with a lathe, and got many useful tips from Threlfall, who was an adept at this work. In spite of Threlfall's attempts to induce the Prime Minister to provide the laboratory, nothing happened for some time, when one evening the Prime Minister came into the Club, went up to Threlfall and said, " Well, Dick, I've done the square thing by you at last ; I've put your laboratory on the Estimates. We've just been beaten on a division and are going out, so the other fellows will have to pay." The result was that Threlfall got a physical laboratory which at the time was one of the finest in the world, and which he soon made into a very active centre of scientific research.

Very soon after I became Professor I was fortunate enough to persuade H. F. Newall (later Professor of Astro-Physics in Cambridge) to come to work in the Laboratory, and for some time we researched together. One of our researches, which led to very pretty experiments, was on the production of vortex rings by drops of one fluid falling through another. When Threlfall went to Australia Newall succeeded him as Assistant Demonstrator. In 1887 W. N. Shaw, who with Glazebrook had organised the teaching of practical physics in the Laboratory and had with him directed the teaching of this subject for seven years, accepted a Tutorship at Emmanuel College. His new duties made too great demands on his time to permit of his continuing his work as Demonstrator, involving as this did attendance at the Laboratory for practically the whole of every other day. He was prevailed upon, however, to continue the courses of lectures on physics which he had given for many years, and he continued to lecture every term until he left Cambridge in 1900 to become Superintendent of the Meteorological Council. He was succeeded as Demonstrator by Newall in 1887, who held office until 1900, when he left the Laboratory to take charge of the large telescope which his father had left to Cambridge University. When Newall was made Demonstrator, L. R. Wilberforce, later Professor of Physics at Liverpool University, succeeded him as Assistant Demonstrator, and when Newall ceased to be Demonstrator Wilberforce succeeded him again. I cannot refrain from quoting an account given by Wilberforce of the relations between them, because it is so true and so honourable to both : " It was impossible to work under him and see not only his ability as a physicist and as an organiser, but also his unfailing patience, kindness and good-humour to his subordinates and his students, without deriving instruction and inspiration from his example " (*History of the Cavendish Laboratory*, p. 255).

Though lectures suitable for candidates for the 1st M.B. Examination were given by Glazebrook, who was lecturer in physics at Trinity College as well as a University Demonstrator, there was no teaching in practical physics suitable for these students until 1887, and no practical work in the examinations. About this time a short oral examination was added as an experiment, and revealed that the knowledge of physics possessed by the candidates was peculiar though not extensive. It furnished the examiners with a crop of "howlers", with which they entertained their friends for long after. In 1887 T. C. Fitzpatrick was appointed Assistant Demonstrator, and I was fortuntate enough to induce him to undertake the organisation of the teaching of practical physics to medical students. At first the classes were small enough to be taken by one man, and did not take up much room in the Laboratory. They grew, however, very quickly, and in about a year

L. R. Wilberforce was associated with Fitzpatrick. The attendance got so large that it was impossible to accommodate the students in the Laboratory, and we had to look out for a new home. We found one at last in a corrugated iron shed which had been used as a dissecting-room. It was not, to begin with, all that could be desired, for such a nauseating odour clung to it that Fitzpatrick and Wilberforce had to exhaust all the resources of physics and chemistry to dispel it. The Laboratory boy was so terrified by fear of the ghosts of the former inhabitants that it was long before he could be persuaded to stay in the room unless someone was with him. When these difficulties had been overcome, the room proved well suited for its purpose. It was big enough to hold large classes, a most important point when the time-table is as crowded as that of a medical student. The classes were held here until a new south wing was added to the Laboratory in 1894. This contained a very large room designed and well equipped for large classes; in this the M.B. classes have been given ever since it was opened.

Glazebrook gave up the lectures to medical students in 1890 on becoming the Principal of University College, Liverpool. The lectures were continued by J. W. Capstick, who had succeeded him as Lecturer in Physics at Trinity College, and when he became Junior Bursar of Trinity in 1898, Fitzpatrick undertook the lectures as well as the demonstrations to medical students; from that time until he became Vice-Chancellor of the University in 1916 he had the entire management of this department of the Laboratory: he made a great success of it, and his work was of outstanding importance to the progress of the Laboratory. This is far from being the only obligation we are under to Fitzpatrick. Of the many gifts which have been received by the Laboratory, none have been more useful than the apparatus for producing liquid air, which he presented to the Laboratory in 1904. Many of the most important researches which have been made in the Laboratory would have been impossible without its aid.

The number of science students in Cambridge increased rapidly after 1885. The pressure on the space in the Laboratory was very great, but it was removed in 1896 when a new wing to the south was opened. The greater part of the cost was borne by the Laboratory, which provided £2000 from the accumulation of fees from students attending the classes. In this wing there is a very large room used for the elementary classes in practical physics, for examinations in practical physics for the Natural Science Tripos and for entrance scholarships to the Colleges. Besides this there was a new lecture-room, cellars for experiments requiring a constant temperature, and a private room for the Professor. The increase in the number of students required an increase in the staff, as we always tried to have one Demonstrator for, at most, twelve students. An additional University Demonstratorship was founded whose salary was paid wholly from Laboratory fees, and S. Skinner was appointed to this in 1891, and C. E. Ashford and W. C. D. Whetham (now Sir William Dampier, F.R.S.), were appointed Assistant Demonstrators about this time. W. N. Shaw, who had been appointed Assistant Director of the Laboratory and had been University Lecturer for thirteen years, left Cambridge in 1900 on his appointment as Secretary to the Meteorological Council. It was decided to abolish the post of Assistant Director and create a new Lectureship, and G. F. C. Searle and

C. T. R. Wilson were appointed to the two Lectureships. Searle only retired in 1935 and C. T. R. Wilson in 1925, when he was elected to the Jacksonian Professorship.

The financial arrangements in force during my Professorship were that the University paid the salary of the Professor ; of two University Lecturers, £50 each ; of two University Demonstrators, £100 each, and of two Assistant Demonstrators, £60 each. The University Lecturers and Demonstrators paid by the University were not nearly sufficient to provide satisfactory teaching for the large number of students in the Laboratory, and it was necessary to provide other Lecturers and Demonstrators whose salaries were paid by the Laboratory. At the end of my tenure this cost about £1500 a year. In addition to the payment of some salaries by the University, the Laboratory received from the Museum and Lecture Rooms Syndicate £270 per annum. The Laboratory had also to pay the wages of the staff of the workshop —and also for the apparatus required for teaching and research, stores and material.

The income of the Laboratory came from fees for lectures and demonstrations, and from the University and Colleges for examinations in practical physics held in the Laboratory.

The distribution of these fees so as to provide for all the varied activities of the Laboratory was left in the hands of the Professor. This method is a rough and ready one, but I think, under the circumstances prevailing at the time, it was the most economical and efficient that could have been adopted. It was the method that had been in force since the foundation of the Laboratory. The fact that the greater part of the income of the Laboratory comes from students' fees is an inducement to make the lectures and demonstrations as attractive as possible, and to start new ones as soon as there is a demand for them. The increase in the Laboratory expenses is not in direct proportion to the number of students. There are some overhead charges which remain the same, and so the profit increases with the number of students. Another advantage is that it is possible with this system to wait until an instrument is wanted before buying it. In the more usual practice, when the University takes the fees and makes a grant to the Laboratory for apparatus, unless you spend the money in the year for which the grant is made, the authorities responsible will think that the grant is greater than you need and reduce it. When the fees are at the disposal of the Professor, he can choose his own time when to spend the money. He may in time accumulate a considerable sum, and my experience is that the most useful instrument a laboratory can possess is a good balance at the bank. You may in this way have at hand a much greater sum than the annual grant, with which you can equip the laboratory with the instruments required to follow up new lines of research, such as radioactivity or positive-ray analysis. The money is ready at hand, and you have not to spend months in getting it from some Committee or Institution. An illustration of this is that, when the work of the Laboratory was suffering severely from overcrowding, a new wing was built for £2000, the result of twelve years' saving at the Laboratory. At that time (1896), the poverty of the University was so great that I was convinced from enquiries that I made that they would not provide this sum. Next with regard to economy. I have gone carefully into the question of what was the cost of the original research going on in the

Laboratory. Taking the accounts for the year 1912–13, the year before the war broke out, and the last in which I was responsible for the research in the Laboratory, I find that an outside estimate for the extra expense incurred by the Laboratory for the researches would be £550, and there were forty research students. The cost of instruments required for the type of research then pursued in the Laboratory was very much less than that of those needed in some of the more recent developments of physics.

In those days the cost of researches done in the Laboratory had to be paid for by the Laboratory itself. The only aid it could get from outside was from the Government grant of £4000 a year for research, which was administered by the Royal Society and had to suffice for the needs of all the sciences, so that there was not much available for any particular science. There has been a great improvement since then. The Royal Society in the last twenty-five years has received very generous donations and bequests, so that in addition to the Government grant it has considerable funds which can be applied to the purchase of apparatus. The Department of Scientific and Industrial Research has made many grants for this purpose; some of the great City Companies and some Colleges, my own among the number, have received bequests which can be applied to aid the researches done by its members. And it was only the other day that Sir Herbert Austin gave the magnificent donation of £250,000 to the funds of the Cavendish Laboratory.

It is not, in general, the first discovery of some quite new physical phenomena that is costly. For example, the discovery of X-rays by Röntgen, of radium by the Curies, the long-continued experiments of C. T. R. Wilson on the formation of drops on particles charged with electricity, cost quite insignificant sums. Discoveries such as these were due to what cannot be bought : to keen powers of observation, to physical insight, to an enthusiasm which does not falter until the difficulties and discrepancies which attend on pioneering work are overcome.

When first the discovery is made, the effects observed are generally very small, and require a succession of lengthy experiments to obtain trustworthy results. It is the attempt to get larger effects that is so expensive. It may mean spending many thousand pounds in making powerful magnets or in producing electromotive forces of hundreds of thousand volts, or in obtaining large supplies of radium. Such money is, however, well spent, for it enables us to obtain new knowledge much more quickly and with greater certainty.

The Cavendish Laboratory was the first of the laboratories in the University where the benefactor defrayed the whole cost of the building, and it remained the only one until well on into the present century, for though a new chemical laboratory and the engineering laboratory had been built before then, the cost was borne wholly by the University. After the war the University was fortunate enough to receive many very large donations, and some of these were for building laboratories. Thus a large physiological laboratory was given to the University by the Drapers' Company. The trustees under the will of Sir William Dunn presented to the University £150,000 for, along with other objects, providing a laboratory for biochemistry and establishing a Professorship in that subject. In 1919 the chemical laboratory received for the extension of the laboratory £210,000 from a number of prominent oil companies. The Rockefeller Foundation provided the University with a very large pathological laboratory ; the Goldsmiths' Company, a laboratory for metallurgy; the Mond Laboratory, for research in magnetism, came from the funds provided by the Royal Society ; and the Low Temperature Laboratory was financed by the Board of Invention and Research. The Institute of Parasitology was created by the munificence of P. A. Molteno. I have mentioned only gifts for laboratories. Since the war twenty-six Professorships in many branches of science and literature have been created by benefactions.

In 1895 the University opened its doors to graduates of other universities whether at home or abroad, and in exceptional cases to those who were not graduates of any university, who wished to come to Cambridge for research or advanced study. These had to satisfy a Committee appointed for the purpose that they were qualified to do research, and that the research they contemplated was one which could be profitably carried out in the laboratory in which they proposed to work. They were entitled to the B.A. degree and certificate for research if, after two years' residence in Cambridge, they submitted to the Committee work which, in its opinion, was " of distinction as a record of original research ". If they had only resided one year when they submitted a thesis which obtained this verdict, they were entitled to the certificate, but not to the degree until they had completed another year's residence. This had such a great effect on the fortunes of the Laboratory, that 1895 is a convenient halting-place in a review of its progress.

After my appointment Glazebrook and Shaw continued their lectures for the first part of the Natural

Sciences Tripos, and for this part I also gave three courses, one in the Michaelmas Term on the properties of matter, and a course on electricity and magnetism extending over the Lent and Easter Terms. For Part II, I gave two courses of lectures, one in the Michaelmas, one in the Lent Term, and in the Easter Term Glazebrook and I set papers for those who were going in for this examination at the end of the term. In addition, I had a class in the differential and integral calculus for students who were hampered in the study of physics by their ignorance of this subject. There were plenty of lectures in the Colleges on it, but they were given by mathematicians for mathematicians, and were not concerned with the applications of mathematics to other subjects. The difference was forcibly expressed by one of my students who was unable to attend my classes, and who went to College lectures. I asked him after about three weeks how he was getting on, and he said he had had to give them up because he had gone to them to learn how to use " Taylor's Theorem " (a fundamental part of the subject), and the Lecturer had talked about nothing but cases where Taylor's Theorem could not be used. Later on these lectures were given by L. R. Wilberforce, and when he left by J. S. E. Townsend (then a research student), now Professor of Physics at Oxford.

Besides attempting to give students of physics some knowledge of mathematics, we organised some demonstrations to enable mathematical students to get some realisation of the physical subjects, heat, mechanics and hydrostatics, included in the Mathematical Tripos. The teaching of mathematical students in physics had been entirely bookish, and many of them had had no opportunities of seeing that their physics corresponded to anything real.

The demonstrations brought to light some interesting points. We found many cases where men who could solve the most complicated problems about lenses, yet when given a lens and asked to find the image of a candle flame, would not know on which side of the lens to look for the image. But perhaps the most interesting point was their intense surprise when any mathematical formula gave the right result. They did not seem to realise it was anything but something for which they had to write out proofs in examination papers. The time-table of the mathematical students in those days was so crowded that it was difficult to find a time when they could attend the demonstrations. The classes were small and were given up after two or three years.

The lectures for Part II of the tripos were much strengthened towards the close of this period by a course on electrolysis by Mr Dampier Whetham (now Sir William Dampier, F.R.S.), who had made important researches on this subject in the Laboratory.

The Cavendish Physical Society was founded in 1893 ; it held fortnightly meetings preceded by tea in term time. Its primary object was the discussion of recently published papers on physics. Besides helping to keep the workers in the Laboratory abreast with recent work in physics, it gave them an opportunity of acquiring practice in lecturing, for the majority of the reports on papers were given by the students themselves. The discussions which followed the reading of the report helped to clear up difficulties, and not infrequently suggested subjects for further investigation. Again, a student who had completed a piece of research generally gave an account of the results he had obtained before publishing them. This was often very helpful, as it enabled him to see the points he had not

made clear, and which would require further explanation
before the paper was published.

During the first two years of my Professorship
Olearski, Natanson and H. F. Reid were working in the
Laboratory. They were the first of a long series of students
from foreign universities who have come to work in the
Laboratory; they all attained distinguished positions as
physicists. It is beyond the scope of this book to give
an account of the many researches that were made in the
Laboratory in this period : I must confine myself to a
small number which contain points of general interest.
H. L. Callendar's career at the Laboratory was in some
respects the most interesting in all my experience. He
was on the classical side when at school and did not do
any physics. As an undergraduate he took a First Class
in classics in 1894 and one in mathematics in June 1895.
He came to work in the Laboratory in the Michaelmas
Term in that year. He had never done any practical work
in physics, nor read any of the theory except in a very
casual way. He had not been in the Laboratory for more
than a few weeks when I saw that he possessed to an
exceptional degree some of the qualifications which make
for success in experimental research. He was a beautiful
manipulator, and delighted in making the results he
obtained as accurate as was possible with the instrument
he was using. The problem was to find a subject for his
research which would give full play to his strong points
and minimise as much as possible his lack of experience.
I knew from the ability he had shown as an undergraduate
that, whatever the subject might be, he would have no
difficulty in mastering the literature about it. It seemed
to me that the most suitable research would be one which

centred on the accurate measurement of electrical resistance. Much work had been done in this subject, and
methods evolved which gave results of great accuracy.
The instrument required—the galvanometer—had been
made so reliable that the use of it did not require a long
apprenticeship. The resistance of a platinum wire depends
upon its temperature, so that if the resistance of the wire
at different temperatures is known you can, by measuring
its resistance, determine its temperature. The wire acts
like a thermometer, and since platinum only melts at a
very high temperature, it can be used for measuring
temperatures at which a mercury thermometer would be
useless. Siemens had actually constructed a thermometer
on this principle, but this was found to have grave defects
which made accurate determinations of temperature impossible. The simplicity and convenience of using a piece
of wire as a thermometer was so great that it seemed to
me very desirable to make experiments to see if the failure
of Siemens' instrument was inherent to the use of platinum
as a measure of temperature, and not to a defect in the
design of the instrument. Callendar took up this problem
with great enthusiasm and showed that, if precautions are
taken to keep the wire free from strain and contamination
from vapours, it makes a thoroughly reliable and very
convenient thermometer. This discovery, which put
thermometry on an entirely new basis, increasing not only
its accuracy at ordinary temperatures, but also extending
this accuracy to temperatures far higher and far lower than
those at which hitherto any measurements at all had been
possible, was made with less than eight months' work. I
had very little to do with it beyond seeing how it was
going on from day to day, and encouraging him when
he was disheartened by the set-backs which occur in all

researches. The results of the investigation were set forth in a paper communicated to the Royal Society in 1886 and published later in their *Transactions*. He was a successful candidate for a Fellowship at Trinity College in the autumn of 1886, and sent in this paper as the dissertation which has to be submitted to the Electors to the Fellowships. The decision of the Electors is influenced far more by the merits of the dissertation than by any other consideration, and Callendar was elected though this was his first attempt. Candidates for Fellowships may try in any of the three years between taking the B.A. degree and being eligible for the M.A., and it is somewhat unusual for a candidate to be elected at his first opportunity. Thus, starting with no previous knowledge of physics, with no experience in making physical measurements, Callendar had in a few months obtained results of absolutely first-rate importance, an importance which cannot be gauged even by their publication in the *Philosophical Transactions* or by their leading to the election to a Fellowship. They gave to the physicist a new tool by which he could determine temperatures with an ease and accuracy never obtainable before. They were also of the utmost importance to industry ; they enabled the steel-maker to measure the temperature of the molten metal in his vat, the brewer the temperature of his brew ; they replaced guess-work by accurate measurement and cookery by science. Callendar's case raises the question whether the very large amount of time devoted in our universities to the performance of experiments by students attending the advanced demonstrations in practical physics is either necessary or desirable. I am inclined to think that it is not, provided the candidate is a good manipulator and is an able man. I think such a one, if he had to use, in the research he was contemplating, some new type of instrument, would not be very long before he had mastered its technique. There have been many great physicists who never attended any demonstrations in practical physics—Joule, Stokes, Kelvin, Rayleigh, Maxwell, to take only English examples—and I am not sure that they lost much by the omission.

Callendar continued his researches on the connection between resistance and temperature at the Cavendish Laboratory until he left it in 1890 to be Professor of Physics at the Royal Holloway College ; he went on with them afterwards in the laboratory at that College, and later still in the laboratory at McGill University, Montreal, where he had been made Professor of Physics. A comprehensive account of these is contained in his paper in the *Philosophical Magazine* for February 1899. In this he says that his experiments suggest that the resistance of certain metals vanishes at a temperature appreciably above the zero of absolute temperature. That this is the case has been proved conclusively by the experiments made by Kamerlingh Onnes and others at the low temperatures which can be obtained by the use of liquid helium and liquid hydrogen.

In addition to his other achievements Callendar, when in Cambridge, invented a new system of shorthand ; he went into it very thoroughly, measuring accurately the time it took to write symbols of different shapes and selecting the fastest. I learnt the system and found it exceptionally easy to read and very useful for making rough notes, as it was not necessary to transcribe them into long hand.

Callendar became a Fellow of the Royal Society in 1894.

W. C. Dampier Whetham (now Sir William Dampier) made important researches in this period with his new method of measuring the velocity of ions in liquid electrolytes, and also on the slipping of liquid when moving in contact with solid surfaces.

Some very notable experiments on the potential difference required to produce electric sparks at different lengths, and through gases at different pressures, between two large parallel electrodes were made by J. B. Peace. He found that when the pressure was kept constant and the distance between the plates gradually diminished, though at first the sparking potential diminished, the diminution ceased when the distance had been reduced to a certain value, and after this the sparking potential increased as the distance diminished, so that it was impossible to get a spark unless the potential exceeded a certain value, about 300 volts in air. Similarly, if the distance is kept constant and the pressure diminished, you cannot get a spark unless the potential difference exceeds 300 volts.

At the end of this period C. T. R. Wilson began to work at the Laboratory and published in 1895 the first of his papers on the formation of clouds, the beginning of a long course of researches which have been of vital importance for the progress of modern physics.

During this period I made experiments on the resistance of electrolytes, and the specific inductive capacity of dielectrics when under the influence of the very rapidly alternating currents produced by the discharge of a Leyden jar. I also made a long series of experiments on the passage of electricity through gases when the electric force producing the discharge was obtained by the forces due to electromagnetic induction, produced when the rapidly alternating currents due to the discharge of a condenser pass through a coil of wire. When a glass bulb containing gas at a low pressure is placed inside the coil, the electro-magnetic induction produces electric forces inside the bulb which are able to produce a discharge through the gas. The discharge passes as a circular ring, since the lines of electric force inside the bulb due to electromagnetic induction are circular. There are in this case no electrodes in the gas ; the discharge is reduced to its simplest form and never has to pass from the gas to another substance. I found these experiments very interesting. The dis-charge is, under proper conditions, bright and very beauti-ful, and the method has valuable applications to the study of many important questions in the discharge of electricity through gases, to the effect on the electrical and magnetic qualities of bodies of very rapid alternations in electric forces acting upon them, and to spectroscopy.

I made, too, experiments on the passage of electricity through hot gases, which supported the view that some kind of dissociation was associated with the conductivity of gas and that the current was carried by charged particles.

The year 1895 is one of the most important years in the history of the Cavendish Laboratory, for then a regulation came into force by which graduates of other universities were admitted to Cambridge as " Research Students ", and if after two years' residence at Cambridge they submitted to a Committee a thesis containing an account of their researches they were entitled to a Cam-bridge degree, provided the Committee declared it was " of distinction as a record of original research ". At first the degree was that of Master of Arts, but after a few years' trial this was replaced by Ph.D. (Doctor of Philo-sophy), a new degree created by the University for their benefit. It was represented to the University that since the M.A. degree did not entitle a man to be called "doctor", our students were at a disadvantage when competing for teaching posts with those who had been to a German university and had obtained the Ph.D. degree. This policy of creating "Research Students" had a very auspicious start, for at the beginning of October 1895, when the regulations first came into force, Ernest Rutherford, now Lord Rutherford, O.M., F.R.S., and J. S. E. Townsend, now F.R.S., and Wykeham Professor of Physics at Oxford, came to the Laboratory to enter as research students within about an hour of each other; and not long afterwards J. A. McClelland, later F.R.S. and Professor of Physics in the University of Ireland until his death. Rutherford, came from Canterbury College, New Zealand; Townsend from Trinity College, Dublin; and McClelland from Queen's College, Galway.

Rutherford began his work at the Laboratory by working at wireless telegraphy, using a detector which he had invented before leaving New Zealand; it worked on the principle that a piece of soft iron wire, magnetised to saturation, loses some of its magnetism when placed inside a solenoid through which the rapidly alternating currents obtained from the wireless pass. He held, not long after he had been at work in the Laboratory, the record for long-distance telegraphy, as he had succeeded in sending messages from the Laboratory to his rooms about three-quarters of a mile away. Townsend began a research on the magnetic properties of the salts of iron of various types, which led to very interesting results. He found that the magnetism due to the same mass of iron was the same in all the ferrous salts and also in all the ferric, but the magnetism in the ferrous was not the same as in the ferric, while in salts like the ferricyanides iron was hardly magnetic at all.

The number of research students increased so rapidly after the introduction of the new regulations that in a few years they became what they have been ever since, a characteristic feature of the Laboratory, giving it quite a cosmopolitan tone. We have had among them students from practically every important university in Europe, Asia, Africa and America. Along with Cambridge students staying on after taking their degrees, there are generally graduates of most universities in Great Britain and the Dominions, some of them holders of the " 1851 Exhibitions", graduates from the United States, and frequently one or more professors of American universities spending their " Sabbatical Year " in research, and graduates from German, French, Russian and Polish universities. The advantage gained by our own students by their intercourse with men of widely different training and experience, of different points of view on political, social and scientific questions, of very different temperaments, can, I think, hardly be overestimated. They gain catholicity of view and some of the advantages they would get by residence in the universities from which the research students came. To give them better opportunities of getting to know each other, I arranged that we should meet for tea in my room every afternoon. Personally I found these meetings very delightful, and I made it a point to arrange my engagements so that I could attend them. We discussed almost every subject under the sun except physics. I did not encourage talking about physics because the meeting was intended as a relaxation for men working all day at physics, and also because the habit of talking " shop " is very easy to acquire but very hard to cure, and if it is not cured the power of taking part in a general conversation may become atrophied for want of use. If the morning papers contained an account of some striking event at home or abroad it was generally mentioned at our tea-party, and very frequently we found that there was someone who knew well the place where the event happened, who was able to throw fresh light upon it and make it seem much more vivid than the reading of any number of telegrams. I have, for example, when a Presidential Election was taking place in the United States, heard Republicans and Democrats fighting their battle with great vigour, and have felt that I learnt far more about American politics by listening to them than by reading columns in the newspapers from special correspondents.

The students from other universities were surprised and at first irritated by the restriction put upon them by the College regulations and particularly by the one which obliged them to be in their rooms before a certain time. No such restriction was in force in any of the universities from which they came, and they said it was like going back to the nursery and being sent to bed at 7 o'clock. Another matter which was a great puzzle to them was the duties performed by their College Tutor. In Cambridge there is a distinction between the office of Tutor and that of Lecturer. The Tutor is supposed to be *in loco parentis* to his pupils, and acts as the connecting link between the College and the parents. He generally holds a Lectureship in some subject, but only those of his pupils who are students of this subject go to his lectures. None of the College Tutors to whom they were assigned ever lectured on physics, and they could not understand what was the use of a Tutor. One of the research students thought he would take a busman's holiday and spend his leisure in a research into what a Tutor did. He proceeded to make experiments. He went to his Tutor and asked him for permission to go to the theatre : the Tutor told him he did not need any such permission. Then after an interval he asked the Tutor to order a supply of coals for the winter to be sent to his lodgings. The Tutor said it was none of his business. He went on like this and at last received his term's bill from the Tutor. He was quite pleased and said, " I have solved my problem : the Tutor sends in the bill ".

By 1898 the number of students had become large enough to justify the celebration of the beginning of the Christmas Vacation by a dinner. This was the first of a series which has gone on without interruption (except in war-time) until now. They are largely attended and highly valued since they enable old research students to meet old friends and talk of old times, and help to establish a connection between past and present research students. The first of these dinners was held in December 1898 at Bruvet's Restaurant in Sidney Street. During the songs after the dinner the proctors came to enquire what the proceedings were about. They did not come into the room where we were dining, being, I suppose, im-pressed, and I have no doubt mystified, by the assurance of the landlord that it was a scientific gathering of research students. A notable feature of these gatherings was the songs specially composed for these occasions by research students. Mr A. A. Robb, the author of a well-known book on the metaphysics of mathematics called *Space and Time*, showed that he could tread a lighter measure with equal success, though I am afraid it requires more mathe-matical knowledge than I can assume to be possessed by all my readers, to appreciate the skill with which he has

turned Maxwell's equations into the chorus of a song to
the tune of " The Interfering Parrot " in *The Geisha* :

d_Y by d_Y less $d\beta$ by dz is equal KdX/dt
While the curl of (X, Y, Z) is the minus d/dt of the vector (a, b, c).

Mr Craig Henderson, who was, I think, present at the
first dinner, was President of the Union in the Lent Term
of 1898, the only Cavendish research student who has
obtained this distinction.

Craig Henderson, while he was in Cambridge, saw a
good deal of the ordinary undergraduate life, a thing
which is not easy for a research student to do. He does
not go to the same lectures as the ordinary student nor,
since he is working in the laboratory all day, can he take
part in their sports : the laboratory takes the place of the
College, but it does not give him as good opportunities
for making friendships with men of different interests, with
different views on social, political, and religious matters, and
with different experiences of life. It does, however, offer
better opportunities of making friends with men of different nationalities. The research students are so enthusiastic
about their work that, besides attending the meetings of
the Cavendish Physical Society, they form societies of
their own, and discuss fortnightly or weekly the recent
advances made in physical science. This, though laudable
in itself, does diminish the opportunities they have for
meeting other than research students, and prevents them
getting from Cambridge all that it could give. It is easy,
however, to exaggerate this, so much depends upon the
man himself. Some are born specialists, others are born
with the gift of making friends and for taking an interest
in a wide range of subjects. And after all, it is not
only the research students who are specialists ; there are
students who specialise in some kind of sport and take

little or no interest in anything else. Others specialise in the Union and take no interest in anything but debates ; others in the cinema, and confine their interests to film stars.

The workers in the Laboratory during the period under consideration, 1896-1900, included L. Blaikie, G. B. Bryan, J. B. B. Burke, W. Craig Henderson, J. Erskine Murray, J. Henry, P. Langevin, J. G. Leathem, R. G. B. Lempfort, Theodore Lyman, J. A. McClelland, J. C. McLennan, C. F. Mott, I. Nabb, H. F. Newall, Vladimir Novak, R. B. Owens, J. Patterson, J. B. Peace, O. W. Richardson, A. A. Robb, W. A. D. Rudge, E. Rutherford, G. F. C. Searle, G. A. Shakespeare, W. N. Shaw, S. Skinner, S. W. J. Smith, Hon. R. J. Strutt, J. Talbot, J. S. E. Townsend, J. H. Vincent, E. B. H. Wade, G. W. Walker, W. C. Dampier Whetham (now Sir William Dampier), R. L. Wills, C. T. R. Wilson, H. A. Wilson, J. Zeleny.

The merits of the research students were soon recognised by the Colleges. Trinity College elected Rutherford to the Coutts Trotter Scholarship in 1898 and Townsend to a Fellowship in 1902. Emmanuel, Caius, St. John's and Trinity have been foremost in encouraging research students and research, and each of these offers each year a studentship open to any graduate of any university other than Cambridge. These, however, are not confined to students of physics.

In the period between the beginning of this century and the war the number of research students steadily increased, and in addition many distinguished physicists came from abroad and spent a year in research at the Laboratory. Among these were Professor Bumpstead from Yale, Professor Carlhelm-Gyllenskold from Stockholm, Professor Erikson from Minnesota, Professor Huff and Professor Mackenzie from Bryn Mawr,

Photo: *Stearn & Sons, Cambridge*

RESEARCH STUDENTS, CAVENDISH LABORATORY, 1898

Left to right, (*Back row*) : S. W. Richardson, J. Henry.

(*Middle row*) : E. B. H. Wade, G. A. Shakespear, C. T. R. Wilson, E. Rutherford, W. Craig-Henderson, J. H. Vincent, G. B. Bryan.

(*Front row*) : J. McClelland, C. Child, P. Langevin, J. J. Thomson, J. Zeleny, R. S. Willows, H. A. Wilson, J. S. E. Townsend.

Professor L. T. More from Cincinnati, Professors Nichols and Hull from Dartmouth College, U.S.A. ; Professor Karl Przibram from Vienna ; Professor Smoluchowski, University of Lemberg ; Professor Vegard, University of Christiania. All arrived in the first decade of the century. They formed a very welcome addition to our society ; they were the most agreeable of companions and we got from them first-hand information of the methods of teaching physics in their own country, and learned which parts they thought from their own experience were satisfactory and which required alteration.

The American Professors were keenly interested in the difference between life at Cambridge and that at an American university. The late Professor Bumpstead, who during the war came to London as liaison officer between our Board of Invention and Research and a similar body in America, to enable each country to become acquainted with the improvements and inventions made by the other, has given his impression of these in a letter published in the *History* of the Laboratory, from which I give a few extracts : " With regard to the University as a whole, one thing which struck me was the essentially British way in which it has utilised survivals of the past, not sweeping them away because they were or might become abuses, but adapting them to modern conditions in such a way that you have a better instrument for the presentation and enlargement of knowledge than we can make here with the ground all clear for the ' most modern improvements '. Two illustrations of what I mean will suffice. One is the division of the University into Colleges, which seems to me to be of enormous advantage to the social life of both undergraduates and dons and to their broad intellectual development. In this country we have schools of engineering, law, medicine and divinity, the liberal arts and so on. The social groups tend to follow these lines with a distinctly narrowing influence. I think you have a much better arrangement and yet I cannot see how it could be made to order. It is only history and tradition that make Trinity, St. John's and Caius all different and yet parts of one whole.

" Another survival which I envy you is the system of Fellowships, which I suppose was the source of some abuse in the past. But I believe it is on the whole the best means of promoting research and sound scholarship which could be devised. . . . If you can keep your Fellowship system from being reformed too much, you will be fortunate among the universities of the world." " Your social life struck me as much richer and fuller than ours, really we seem to work too much to enjoy each other's society and yet we do not get so much done as you. I don't understand it altogether : a laboratory in this country in which nobody ever began work before 10 A.M. or worked later than 6 in the evening would serve as a terrible example of sloth and indolence. I do not see how you can get so much work done and yet have time to live so pleasantly and unhurriedly." "I have never seen a laboratory in which there seemed to be so much independence and so little restraint on the man with ideas."

The election of a man of science to the Fellowship of the Royal Society of London, the most important scientific society in this country, is a proof that his work has been of exceptional importance and merit. This honour has been conferred upon many of those who worked in the Cavendish Laboratory while I was Professor. See Appendix.

A large number of students who worked at the Cavendish Laboratory whilst I was Professor were elected to Professorships in Physics in this and other countries. The list of universities in which my pupils have held Professorships is given in the appendix; when more than one Cavendish student has held the Professorship, the number is enclosed in brackets.

In 1902 a portrait of myself by Mr Arthur Hacker, R.A., was presented to the Laboratory by those who had been, or were at the time, my students. I think it is the portrait I like best. It was painted in very few sittings, as I had to sail to give some lectures at Yale, and there were very few dates which suited both the artist and myself. This is an example of what has always been my experience, that a few sittings give better results than a great many. I have had twenty-five and twenty-seven sittings for other portraits, and have watched the picture gradually getting worse after each sitting. Perhaps the most successful of all was a bust by Mr Derwent Wood, R.A., now in the library of Trinity College, for which I only sat for about forty-five minutes.

The portrait by Mr Hacker hangs in the Laboratory. Another example of the generosity and goodwill of my old students came on my seventieth birthday when they gave me an address with 259 signatures, together with a large cigarette-box, which I am afraid I use too often. These were presented to me at a dinner, when speeches were made by several very old friends and pupils : Lord Rutherford, Professor Langevin, Dr. A. Wood, Sir Arthur Schuster, Sir Richard Threlfall and Dr. Horton. A charming present was also made to my wife, who was at the dinner. These expressions of goodwill touched me very acutely ; the remembrance of them is delightful. They recall the help, kindness and goodwill which it has been my good fortune to receive from everyone connected with the Laboratory, for more than fifty years. I owe to the Laboratory not merely the opportunities it has given me for indulging my scientific tastes in a way that would not have been possible in any other place. I owe to it besides many valued friends, and memories of friendships which can no longer be more than memories.

CHAPTER V

Psychical Research

IN the nineties, at the instance of F. W. Myers, I attended a considerable number of séances at which abnormal physical effects were supposed to be produced. This was the only kind I went to see. I did not attend any demonstrations of thought transference, or those where the medium showed a knowledge of one's personal affairs which she could not have acquired by natural means. The results were very disappointing; at all but two of those I attended nothing whatever happened, and in the two where something did there were very strong reasons for suspecting fraud. The first of these was given by an American slate-writer, Eglinton, whose séances had received a good deal of attention from the Press. He claimed to be able to get messages written on slates under circumstances which precluded human agency. I went with Myers and Mr H. J. Hood, who at that time took a prominent part in psychical research, to his room, near, I think, to the Marble Arch. Eglinton took a slate which we were allowed to examine, and we found no reason to suspect that it was anything but an ordinary school slate. He then broke a small piece off a slate pencil, and placed the fragment on the top of the slate. We then sat down at a trestle table ; he sat at one end, I held his right hand with my left, and with his left hand he held the slate under the table. The piece of slate pencil was between the bottom of the table and the top of the slate. The room was not darkened in any way and all the proceedings took place in broad daylight. He then asked each of us for a question which we should like the spirits to answer. Myers and Hood asked questions concerning spiritualism, I preferred a question to which I knew the answer, so I asked what county Manchester was in. We then sat for, I should think, a quarter of an hour without anything happening. Then he seemed to be seized with convulsions and it was all I could do to hold him up and prevent him from falling off the chair. He recovered in a short time, brought the slate from under the table, and on it was written in a sprawling hand with large ill-formed letters : *Manchester*. My view is that Eglinton thought there must be some catch in my very simple question, and that he knew that some English towns were counties in themselves and supposed, because I had asked the question, that Manchester was one of them. With regard to the way in which the writing was done ; it is possible to write on a slate either in the usual way by keeping the slate fixed and moving the pencil, or, though with much greater difficulty, by keeping the pencil fixed and moving the slate. Thus, if he had managed to jam the piece of slate pencil in a crevice or depression on the lower part of the table, he might have been able to write without any considerable movement of the arm supporting the slate.

A much more exciting and interesting experience was one with Eusapia Palladino, an Italian peasant woman who had been discovered by Professor Richet, a celebrated French physiologist, and who had shown very remarkable powers of producing abnormal physical effects in a series of séances she had with Professor Richet and Sir Oliver Lodge in an island in the Mediterranean under conditions which made fraud seem very improbable. Eusapia came to Cambridge in the Long Vacation of 1895, and stayed with the Myers. She held some séances and I was present at two of them. At the first, which began about 6 o'clock in the evening, Lord Rayleigh, Professor Richet, Myers, Richard Hodgson, Mrs Sidgwick, Mrs Verrall and myself were present, along with a few others whose names I have forgotten. We sat at a long table. Eusapia was at one end, Lord Rayleigh on her right, I on her left ; and Mrs Verrall was under the table holding her feet. There was a melon on a small table at some little distance from that at which we sat, and it was part of the programme that the melon should be precipitated on to the table. There were heavy velvet curtains over the windows, and when the lights were all put out it was pitch dark. We formed the circuit by clasping hands in the usual way, and sat like this for a considerable time without anything happening. Then Myers, who thought it good policy to encourage mediums at the commencement of a séance, jumped up and said he had been hit in the ribs. The circuit was thus broken and it was perhaps a minute before it was re-formed. Hardly had we got seated when Mrs Verrall called out " She's pulled her foot back ", and then, without an interval, "Why, here's the melon on my head!" What had happened was quite obvious : while the circuit had been broken and Eusapia was free she had reached out, got the melon, sat down and put it on her lap, intending to kick it from her knee on to the table. Now if you want to kick from the knee you begin by drawing the foot back. Eusapia had done this, but had been disconcerted by Mrs Verrall calling out, and had not got her kick in in time, and the melon had rolled off her lap on to Mrs Verrall's head. She then set to work to retrieve the situation.

Very soon after we had sat down again she began abusing me in a language of which I did not understand one word, but Richet, who understood her, said she was accusing me of squeezing her hand and that she would not allow this. Instead of holding her hand, all she would permit was that I might put the tips of my fingers on the back of her hand; this was done. After a short interval she began to abuse Lord Rayleigh, and Richet said she was accusing him of squeezing her hand and that, instead of holding hands, she would put the tips of her fingers on the back of his hand. This change, as was found out later by Mr Hodgson, was the essence of her trick. She kept moving the hand I was touching backwards and forwards and I had to follow with my fingers. In this way she was able to make my fingers slip from the back of her hand either to her other hand or on to the back of Lord Rayleigh's. In this case she would get both hands free, for I should think that I was pressing Eusapia's hand when as a matter of fact I was pressing Rayleigh's, and he would think he was still being pressed by Eusapia while in fact he was pressed by me. If she slipped my fingers to the back of her other hand and not to Rayleigh's she would get one hand free. Certainly, soon after this change in the way of holding hands was made, the proceedings became very lively, chairs came down on the table with a bang, the table rose in the air; at any rate the end near me did. We were hit on the back and dug in the ribs, and this went on for what seemed several minutes. After this was over she showed us another phenomenon which, unlike the first, was done with a light in the room. It was not a very bright light, but good enough to enable us to distinguish people and the position of the furniture. There was a window at one end of the room covered with a heavy velvet curtain. She sat down on a chair a few feet away from this with her back towards it. She made no objection to my looking between the curtain and the chair to see if there was a string or connection of any kind between them. I did not detect any, but the light was not good enough to enable me to be certain that nothing was there. After we had been seated for a short time the curtain began to move and, looking like a sail full of wind, reached Eusapia and then slowly fell back. It did not look as if it had been pulled by a string, at any rate by a single string, for if so it would have been pointed like a jelly bag instead of being round like a sail. As she was to give another séance on the next evening, I went to the house in the morning and asked Myers if I might fit up an arrangement to make sure there was nothing between Eusapia and the curtain. He agreed at once. I fastened a string to the wall on one side of the window ; it fell at first straight down, then passed along the floor between the curtain and where Eusapia sat, then went up to a pulley near the top of the wall on the other side of the window, and then to a place where the note-taker sat. The arrangement was that at a given signal the note-taker would pull in the string and if there was anything between the curtain and Eusapia it would be caught by the cord. The arrangement was not at all conspicuous unless you were close to it. This experiment proved a failure, for when we were assembled the next evening at the other end of the room before the lights were put out, Eusapia when she came in became very angry, went up to the curtain and pulled the string down. I have no satisfactory explanation of the way the curtain was moved, but I only saw it on one occasion and that for but two or three minutes, so that it is likely that I omitted to notice something which might have given the clue.

It ought to be said that Sir Oliver Lodge, Professor Richet and Dr. and Mrs Sidgwick observed Eusapia under much more favourable conditions at Professor Richet's house in an island near Hyères, and that Sir Oliver is still of opinion that the effects observed there were genuine, and that the failure at Cambridge later was due to her having lost most of her power in the meantime and supplanting it by conjuring. The phenomenon of the bulging-out of curtains without any apparent connection with Eusapia, and when there was no wind also, occurred on the island. The explanation favoured by Lodge is that she had the power of thrusting out a sort of supplementary arm, which has mechanical qualities of pushing or pulling but which is not ordinary flesh and blood. This substance has been named ectoplasm. Lodge, on at least one occasion, thought he could see in the faint light a protuberance stretching out from Eusapia. It looked like a mere stump and not a hand. Within the last few years an Austrian medium, Rudi Schneider, has produced when in a trance phenomena similar to Eusapia's, including the bulging of curtains. Attempts have been made both in Paris and London to photograph the ectoplasm using the invisible infra-red rays, which can be used in the darkened room where these phenomena are supposed to occur. These experiments, however, have not led to any conclusive results. In the experiment in Paris it was thought that the ectoplasm produced an appreciable absorption of the infra-red rays, but in those in London, where infra-red photographs were taken with an apparatus designed by Lord Rayleigh, no effects which could be ascribed to ectoplasm were observed. I think that though physical instruments of very great delicacy have been used to detect physical effects produced by psychical means, they have as yet not given any evidence of their existence. Perhaps this is hardly to be wondered at. The people who claim to produce them are very psychic and impressionable, and it may be as unreasonable to expect them to produce their effects when surrounded by men of science armed with delicate instruments, as it would for a poet to be expected to produce a poem while in the presence of a Committee of the British Academy.

Again, the delicate instruments used in physical laboratories may, until their technique has been mastered, give one result one day and a contradictory one the next, and illustrate the truth of Coutts Trotter's saying that the law of the constancy of Nature was never learned in a physical laboratory. The most complicated physical apparatus is simplicity itself compared with a human being. I have described the only two cases in which anything happened that was claimed to be psychical. One of my most interesting experiences was a séance when nothing at all happened. This was given in Mr Oscar Browning's rooms in King's College, Cambridge, near the end of last century by Madame Blavatsky, a very prominent theosophist of the time. She said at the beginning that her Mahatma in Tibet would precipitate a message, a cushion and a bell, and we sat waiting for, I should think, more than an hour, and nothing whatever arrived. The medium was not in the least abashed. She took the offensive, said it was all our fault, that our scepticism had created an atmosphere impenetrable to anything spiritual. She was a short and stout woman with an amazingly strong personality, very able and an excellent speaker. So well did she speak that she convinced the great majority of the audience that the failure was their fault, and they went away thoroughly ashamed of themselves for having spoiled what would otherwise have been a most interesting experience.

Telepathy

Another branch of psychical research which may be connected with physics is that of thought-reading or telepathy, especially that between people not far from each other. The first experiments published on this subject were made in 1884 on girls employed in a draper's shop in Liverpool. These were tested in the physical laboratory at Liverpool University by Sir Oliver Lodge, and he satisfied himself that they, when blindfolded and when precautions were taken against the use of a code of signals, could describe, and in some cases give a rough copy, of a drawing made in sight of the rest of the audience and therefore presumably in their thoughts. Compared with other branches of psychical research, little has been done on this short-range thought transference between living people, though it was this which first suggested the idea. One reason for this is that this power of thought reading is exceedingly rare, very much rarer than was at first supposed. Another reason is that attention was at first directed to thought transference between the living and the dead, which raises much deeper and more important questions. In my opinion the investigation of short-range thought transference is of the highest importance. It is quite possible, indeed very probable, that it may turn out to be of an entirely different character from the kind of thought transference that is supposed to occur in dreams or premonitions. This does not seem to be influenced by distance, whereas one would be very much surprised if that between living people did not decrease rapidly as the distance between them increased. I think it is now recognised that it is much more difficult than was then thought to guard against the use of a code. The Zancigs, who were performing in London some years ago, did not, I believe, profess to possess any psychic power. Yet I saw them at a private house give, under most stringent conditions imposed on them by the present Lord Rayleigh (conditions which I expect were even more stringent than those imposed in the experiment of fifty years ago), an exhibition which, if they had professed to be thought readers, might have been held to establish the claim.

There are other ways of communication between two people than by sight or audible sounds. There is such a thing as inaudible sound. If the pitch of a musical note is raised above a certain point it ceases to be audible. The pitch at which this occurs varies from one person to another, and some people are able to hear notes of a pitch so high that they are quite inaudible to the vast majority of people. I once saw a very striking instance of this at the Cavendish Laboratory. A boy, who is now a distinguished musician, came to the Laboratory to have the limiting pitch he could hear tested by Galton's whistle. This is a whistle whose pitch can be altered continuously through a wide range. He beat the instrument, for we could not produce a note of high enough pitch to be beyond his hearing. There were several people present, some of them young students—and the power of detecting high notes is much stronger in youth than in age—and none of them had heard anything for a long time before the whistle reached its highest note. Now if one of the parties in a performance like the Zancigs possessed this power to a quite exceptional extent, all that the other one would need to communicate with her by a code would be a very small whistle which could easily be concealed. The chances of anyone else hearing the signal would be small. I do not for a moment suggest that the Zancigs did the trick in this way. I have no idea how they did it.

In my opinion by far the most important experiments of this kind were those made by Professor Gilbert Murray in 1916. In these there could be no question of the use of a code or of the good faith of any of the participants. An account of these is given by Mrs Sidgwick in *Proceedings of the Society for Psychical Research*, vol. 34, p. 212. The participants in these experiments were either members of Professor Murray's family or intimate friends. The experiments were made in private houses and not in a laboratory. This has the advantage that the thought reader was working under conditions much less formal and much more normal, and less likely to affect his powers, than if the experiments had been made in a laboratory.[1] Professor Murray says the method followed is this : " I go out of the room and of course out of earshot. Someone in the room thinks of a scene or an incident or anything she likes and says it aloud. It is written down and I am called. I come in, usually take my daughter's hand, and then if I have luck, describe in detail what she has thought of. The least disturbance of our customary method, change of time or place, presence of strangers, controversy and especially noise is apt to make things go wrong."

[1] It is perhaps pertinent to remark that the original experiments in Liverpool, which were regarded as proving as well as suggesting the idea of thought transference, were made in a laboratory.

Professor Murray himself does not believe that the results he gets are due to telepathy, if by that is meant the transmission of thought without aid from the ordinary senses, hearing, sight, touch, etc. As I understand him, he believes he gets into a state when he is peculiarly sensitive to noise, and his hearing becomes so acute that he hears something of the conversation between the "thinkers" when they are settling the subject they are to think about. It is not so acute, however, that he can distinguish the words they say, but he hears enough to suggest something to him. If it were a question of acuteness of hearing alone a powerful microphone might be as effective as Professor Murray. The process is easier to describe, and the principle is the same, if we suppose the subject was chosen by writing instead of talking, and that the thought reader was in a condition when his sight was abnormally acute, though not sufficiently so to enable him to fix definitely what a particular word was. It might, however, be sufficient for him to see that only a few words would be consistent with what he had seen. If he had a quick imagination he might quickly grasp these possibilities and, beginning with one of them, if he found from observation of the thinkers that he was on the wrong track, start one of the others and go on until he got the right one.

It is clear that if this were the process the percipient would have a much better chance of success if, as in this case, the thinkers were either members of his family or intimate friends with whose special interests in literature, history or politics, as well as incidents in their lives, he was well acquainted.

The subjects selected by the thinkers seem to have been chosen at random, without reference to their suitability for distinguishing between the psychic explanation and the one favoured by Professor Murray. There are some of the experiments which seem, at first sight at any rate, to be inconsistent with his view, but it would be possible to find subjects which would test this view more severely than those actually used.

Many efforts have been made to discover possessors of this power of short-distance telepathy, both in this country and in America, but they seem to be exceedingly rare. The case of Professor Murray shows, however, that few though they may be, there are some, and it is in my opinion most important that the search for those should not be dropped. It is often asserted that telepathy has been conclusively proved. I cannot agree with this for the case of short-range telepathy. I think the utmost that can be claimed is the Scottish verdict— not proven. This does not mean that it has been shown not to exist. The subject is one of transcendent importance. I agree with the late Lord Rayleigh, who in his Presidential Address to the Society for Psychical Research, given very shortly before his death, said, " To my mind telepathy with the dead would present comparatively little difficulty when it is admitted as regards the living. If the apparatus of the senses is not used in one case, why should it be needed in the other ? "

Water-Dowsing

The phenomenon of water-dowsing, which borders on the occult, would seem at first sight one which could be more easily investigated than thought transference between living people. For there are a considerable number of people—in fact there is a Society of Dowsers—whose good faith is beyond question, who suffer a strange experience when they walk over ground under which a stream of water is flowing. The most usual form of this is that when they hold a forked twig, generally a hazel one, with one branch of the fork in the right hand and the other in the left, the straight part is violently deflected when they are passing over running water. It seems necessary that the water should be underground, for few, if any, cases are recorded of this sensation being felt when crossing a bridge over a running stream. Some dowsers say that the twig will move when held by someone else provided they place their hands on his shoulders. There are some who can detect water without using the rod, since they experience peculiar sensations when they are in places where the rod would move. The rod may be a convenient indicator of a physiological effect. There is no doubt of the reality of the dowsing effect. In fact, in many agricultural districts the dowser is the man they call in when they want to find the right place to dig a well, and he very often succeeds. We had an example of this at Trinity College. The water supply to one of our farms was very defective and a new well badly wanted. At first the Senior Bursar, who was a Fellow of the Royal Society, proceeded in the orthodox way and employed eminent geologists to report on where we ought to sink a well. Their advice, however, did not lead to the discovery of any water. Our land agent said, " If I were you, I would try old X, who has found a good many wells in this county and who will sink the well on the terms ' no water, no pay ' ". As there seemed nothing else to be done, the Bursar employed him and he found water. For this we were assailed in an article in *Nature*, which lamented that Trinity College—the College

of Newton—should have given countenance to such super-
stitious and unscientific practices.

Although I think most of the people who have paid any
attention to the subject believe in the reality of dowsing,
there is no agreement about its cause.

The common-sense view would be that, since underground water would produce some effect on the vegeta-
tion on the surface above it, a man who had good powers
of observation and long experience, might be able to detect
water by small differences in the herbage which would
escape the notice of the ordinary observer, and that the
movement of the rod came after the discovery and not
before. This was the method of the Abbé Paramelle, a
most successful discoverer of springs, who claims that in
the twenty-five years between 1829 and 1854 he had located
more than 10,000 springs; and that between 8000 and 9000
wells had been dug of which over 95 per cent were successful. He did not use the divining-rod and spent all his
time, except on Sundays, looking for water. This may be
the method in some cases but it certainly is not so in all.
In some cases the movements of the rod occur when the
dowser has been blindfolded, or when he has not been near
the part of the country before. In this connection it may
be pertinent to say that in the early days of radio-activity
at the beginning of this century, I examined specimens of
water freshly drawn from a good many wells in different
parts of England and found that they all, though in very
different degrees, contained the radio-active emanation
from radium. This only retains its activity for four days,
so that if the water were stagnant it would soon lose its
radio-activity. Thus, if this were connected with the effect
produced on the dowser, it would explain why it is that
he can only detect *running* water. If inhaling a small

quantity of radium emanation produced in the dowser rigors such as are produced by some drugs, his fingers when over the water might close with a convulsive clutch on the forks of the twig and require a large force to keep the rod in position. I must, however, confess that though I have often inhaled the emanation I have never observed that it produced the slightest effect upon me ; but I am not a dowser.

One amusing thing about the tests of the waters from these wells was that by far the most radio-active was that from a well-known brewery. When I reported this to them I at once received instructions that I was not on any account to publish it, for if I did nobody would ever buy their beer. When I told this to an American friend he said it illustrated the difference between the business methods in the two countries, for if it had been an American brewery, the whole country would have been placarded with advertisements—" Drink Smith's Radio Beer ".

The early history of the " divining-rod " does not inspire us with much confidence. Its first applications were to the detection of criminals, metals and lost treasure and not to the discovery of water. Aymar, a French dowser, created a great sensation in 1692 by tracking down by the divining-rod the murderer of a man and his wife. He claimed that when he placed his foot on anything which had been touched by the criminal the rod moved, his temperature rose, his heart beat more rapidly and he felt faint. These sensations were only produced by murderers ; thieves, metals, or water left him unmoved. The divining-rod was used to discover metals in Germany (Dousterswivel in *The Antiquary* was a German), Austria and Cornwall. In this country attention has been mainly concentrated on water-divining, though some of the most successful water-diviners have claimed to be able to detect metals as well. Thus it is stated that John Mullins, a very celebrated water-diviner and well-sinker, when a sovereign had been placed under each of three stones in a row of ten in a road, passed his rod over the stones. When he came to a stone and the rod did not move he said, "Nothing here", but if it did move, he lifted the stone, took the sovereign and said, "Thankee, Master". This is an experiment that could easily be tried. There are some dowsers who claim to be able not merely to detect metals but to be able to tell what the metal is. The most unexpected application of the divining-rod I have come across is a medical one. I have met a dowser who told me that, before taking a bottle of medicine, he passed the rod over it and if the rod were deflected he knew the medicine would do him good and he took it, while if the rod were not deflected he knew the medicine would be of no use for his complaint.

The greater number of experiments with the divining-rod have been made for the purpose of obtaining water and not for discovering the way in which the effects of the rod are produced. The object has been commercial rather than scientific.

In Barrett's *Divining Rod*, p. 256, an account is made of an experiment made to detect hidden coins by the divining-rod, which was successful, though there were one or two incidents which might arouse suspicion.

The divining-rod is perhaps of all phenomena which may be thought to be psychical, the one most favourable for experiment. The motion of the rod is a mechanical effect and gives an indication of the magnitude of the phenomenon. The conditions under which the effect occurs can be made definite, and there is no lack of trustworthy people who possess the dowsing power : for these reasons I think such experiments are well worth making.[1]

[1] Since the above was written, I have seen a paper by Mr H. M. Budgett, published by the British Society of Dowsers, on the connection between effects on a divining rod and the readings of an instrument designed to detect very penetrating radiation. There seemed to be indications of some connection, but these were very faint. It would be interesting to see if the dowser could, by his divining, detect a strong beam of penetrating γ rays.

CHAPTER VI

First and Second Visits to America.   1896, 1903.

IN the autumn of 1896 my wife and I went to America
to attend the celebration of the Sesquicentenary of
Princeton University, at which I was to give a course of
lectures on the Conduction of Electricity through Gases.
We crossed in the *Campania*, then quite a new boat.
When we reached New York there was some difficulty
in getting our baggage through the customs. I had a
suitcase full of congratulatory addresses from various universities and scientific societies in England. Each of these
was in a cardboard cylinder. There had been some rioting
and bomb-throwing about this time in America and the
inspector at the Customs at once suspected bombs. He
asked me what these cylinders were, and I said they contained addresses to Princeton University. He then asked
why I was taking them. I said it was for their Sesquicentenary. " Oh ! " he said, " I see, you're a foreigner.
But where's Princeton anyhow ? " I said it was quite
close to New York. He said he had never heard of such a
place, and appealed to another inspector, who said he had
not either. He refused to let my baggage through. I then
demanded to be taken to the head man. After some more
talking he consented, and took me to his office. Very
fortunately he turned out to be a Princeton man, who
received me with enthusiasm when he knew my errand
and was not at all pleased with his subordinates for what
seemed to him incredible ignorance.

FIRST VISIT TO AMERICA

The morning after our arrival we went to Baltimore to stay with some great friends of ours who lived there. We were fortunate enough to arrive just in time for me to see the final round in a great baseball competition. One of my illusions was soon dispelled. I had always thought Americans as a race would show the characteristics of their ancestors who came over in the *Mayflower*, would be grave, undemonstrative, and would take their pleasures sadly. But though I was in one of the best places and surrounded by the most prominent men in the city—lawyers, bankers, doctors, professors and even clergymen—the game had only been going on for a few minutes when most of these were shouting at the top of their voices ; some of them shouted so much that they lost the power of articulation and could only croak. I always myself get very much excited by a keen contest and feel for the moment that nothing on earth matters so much as that the side I am interested in should win, and probably if I had wished very much that one of the sides should beat the other and had understood the finer points of the game, I should have done as they did. I enjoyed the match thoroughly. I had never seen baseball before and it is a great game. The idea is the same as that of the English game of rounders in that the batsman, armed with a broomstick, tries to hit the ball so far that he has time to run to a base before a fieldsman can gather the ball, throw it at him and hit him ; but there is as much difference between this and a good baseball match as between cricket played by boys in a side street with three white lines on a wall for the wicket, and the match between Gentlemen and Players at Lord's. The fielding, *i.e.* the throwing and catching of a first-class baseball team, is superb and reaches a standard but rarely attained in English cricket. The pitcher, who corre-

sponds to the bowler at cricket, throws, not bowls, the ball
full pitch at the batsman, and by putting spin on the ball
can make it swerve in the air either to the right or to the
left, or up or down. This interests the physicist as well as
the sportsman, for it is a striking illustration of the hydrodynamical principle that a ball moving through air always
tends to follow its nose, the nose being the point on the
ball which is in front. Thus if a ball is moving horizontally in the direction of the arrow, the foremost point on
the ball will be N, the point right in front of the centre C.
If the ball is not spinning, this point, like every other part

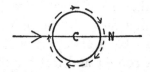

of the ball, is moving horizontally, and the ball, following
its nose, will do so too. But suppose the ball is spinning, as
in the figure, about an axis at right angles to the plane of the
paper, N, in consequence of the spin, will be moving downwards (this spin is what is called top spin in lawn tennis)
and the ball will tend to follow it and will therefore dip ;
if the spin were in the opposite direction, the nose would
be moving upwards so that the ball would tend to soar.
If the axis of spin were vertical instead of horizontal,
the nose N would have a sideways motion and the ball
would swerve either to right or left. If the axis of the
spin were not truly vertical or horizontal, the ball would
both swerve and dip or rise.

We saw Baltimore under exceptionally favourable
circumstances, for our host was a professor in Johns Hopkins University, and both he and his wife belonged to old

Baltimore families. We thus saw much both of academic society and also of that representing the old social life of Maryland. I have never been in a city where the hospitality was more charming, more generous or more intimate than that of Baltimore. Everyone seemed to know everybody else and to call him by his Christian name, and generally omit the other. The result was that I knew the Christian names of a great many people but very few of their surnames. This was especially the case at the Maryland Club, known all over the world for the excellence of its cooking ; indeed it has been called the " gastronomic centre " of the universe, and certainly I have never been in one where the cooking and the food were better. The standard of cooking in private houses is also very high. It is here that you get that excellent dish, Maryland chicken, at its best.

The country round Baltimore is the epicure's paradise. The shores of the Chesapeake Bay supply abundance of oysters of many kinds. In a restaurant at which I had lunch, two pages of the bill of fare were occupied by the names of different kinds, some of these as large as those which Thackeray ate when he was in America, and which he said made him feel that he was eating a baby. The Bay too, is the home of the celebrated delicacy, the canvas-back duck, which plucks the celery from the bottom of the marshes while another duck, the red head, waits until the canvas-back comes to the surface and tries to rob him of the celery. This dishonesty seems to affect the flavour, for though the ducks are much alike and eat the same food, I have heard a Baltimorian, when he wished to describe adequately the monstrous stupidity of someone under discussion, say that he was so stupid that he could not tell the difference between a canvas-back and a red head.

I should think Baltimore must be a paradise for young girls, for the " débutantes ", *i.e.* the girls who have " come out " not more than a year ago, are the Queens of Society. It was for them most of the parties seemed to be given. Their parents placed the drawing-rooms at their disposal to entertain any guests they liked. For one year they reign, then they make way for the next crop.

Our hosts had a most delightful example of the old negro servant who had been in the family for many years, even before the liberation of the slaves. He took a keen personal interest in the guests, would advise them about what they should do and where they should go and whispered at dinner the wine he would recommend. His manners were so good that all this seemed quite natural and very agreeable. He had refused to leave when he had been liberated after the Civil War, saying that he had no opinion of free niggers. It takes some time before an Englishman gets to realise the feeling of Americans towards the coloured people. I was helped by the following incident. A kindly old gentleman was talking about Booker Washington, the coloured statesman who did wonders for the improvement of his race. He said, " I have always been accustomed to call a coloured man ' Uncle ', but I felt it could not be right to call such a great man that. I could not bring myself to call any coloured man ' Mister ', and I did not know what to do. At last I had a bright idea, and I called him ' Professor '."

Johns Hopkins University in Baltimore was, I believe, the first university to be founded where the primary consideration was research and not teaching. There were at first no undergraduates—the students had already graduated at some other university. The first President was Professor Gilman, and he had the very difficult task of selecting a staff of Professors suitable for this purpose. Among the earlier Professors were Henry Rowland, the great spectroscopist, whose investigations with diffraction gratings made in Johns Hopkins laboratory started a new era in spectroscopy ; he spent most of the later years of his life in inventing and developing a printing telegraph, one which, like many now in use, would print the message as it was received (in this he was before his time and I am afraid did not gain much by it) ; Ira Remsen, Professor of Chemistry, who later became President; G. L. Gildersleeve, Professor of Greek.

Among the early students was Woodrow Wilson, subsequently President of the United States, who studied history for a year.

The new university had to fight against several disadvantages. Johns Hopkins had not been a popular citizen and people were reluctant to give money to a university called after him : again, most of the professors and students were northerners while Baltimore was distinctly southern, and in 1876 the feeling aroused by the war was still strong enough to make them unpopular. Indeed, though Johns Hopkins has been the pioneer of the " Research University " not only in the United States but in the world, it has never received, as it deserved to receive, benefactions comparable with those received by some other American universities. It now, I believe, does admit undergraduates, and by the aid of the State has been able to obtain adequate buildings on a fine site. I believe that when Johns Hopkins University first applied to the State for a grant, their representative (a Professor) had to go before a committee of town councillors. One town councillor was very anxious to find out how many hours a day a Professor spent in teaching, and asked the Professor this question. The Professor said it was a difficult question to answer, there were so many things to be taken into consideration. He was asked, " Does it take four hours a day ? " He said, " Perhaps about that ". " What do you do the rest of the day ? " he was asked. " Well," he said, " some time was taken up in the preparation of his lectures." " What ! do you mean to say you do not know your lessons yet ? If one of our school marms did not know her lessons, we should soon get rid of her."

The preparations for the Presidential Election were in full swing when we were in Baltimore. It turned on the question of bimetallism and excited a quite exceptional amount of interest. The bimetallic candidate was Bryan, who was thought to have owed his nomination to his phrase " Thou shalt not crucify humanity on a cross of gold ", which ran like wildfire through the country. McKinley was the other candidate, and he supported the maintenance of the gold standard. I never knew an election excite such bitter feelings. The language of some highly respectable people gave one the impression that, if they could have killed Bryan without any danger of being found out, they would have done so without a moment's hesitation.

After a most pleasant but too short stay in Baltimore we went on to Princeton for the delivery of my lectures, staying with Professor and Mrs Fine. Fine was the Professor of Mathematics and was a great friend of Woodrow Wilson, who, after he became President, wished to make him American Ambassador to Berlin. Fine, however, had to refuse it on account of the expense it would have entailed.

My lectures were very well attended and I was able to meet many of my audience under conditions more favourable for social intercourse than lectures. Princeton, of all the American universities I have visited, is the one most reminiscent of Cambridge. It is not surrounded by workshops or shops, but by spacious lawns studded with fine trees which recall the Cambridge " backs ". To a Trinity man, especially, it recalls Cambridge, for the gateway to one of the courts is copied from the Great Gate at Trinity. When I last visited Princeton after an interval of more than twenty years I found the resemblance still greater, for they had erected a beautiful building for the post-graduate school. I had the pleasure of dining there one evening, and, just as in Trinity, we dined in gowns and had a Latin grace, and if I remember correctly, port after dinner. The dinner was the more agreeable since Dean West, to whose exertions the school is mainly due, was in the chair.

The houses of many University Clubs along Prospect Avenue are a very interesting feature of Princeton. They are very attractive buildings and luxuriously furnished, though I was told that they were so well endowed through the munificence of old members that the charges to the undergraduate members were not extravagant. Those Clubs have played an important part in American history, for it was owing to them that Woodrow Wilson became President of the United States. It came about in this way: to be or have been a member of some of the Clubs gave a man considerable social distinction, not only among his fellow undergraduates when he was at Princeton, but also for the rest of his life in most of the eastern States. The result was that some of the undergraduates thought that the most important thing they could do at Princeton was to " make ", *i.e.* get elected to, a good Club like, say, " Ivy ". They could not be elected until the end of their second year, so that their first two years were devoted to doing the things which they thought would best promote their election : unfortunately they believed that it was fatal to their chances if they did any work. This clearly called for reform, and Woodrow Wilson, when he was made President of Princeton in 1906, set to work to remedy it. He began by raising the standard required for entrance to the University, and supplemented the lectures of the Professors by the appointment of preceptors, who give individual attention to the students. Mere attendance at lectures is a very unsatisfactory method of teaching for any but the best student. The ordinary student needs someone who will explain his difficulties, set him essays and discuss them with him. This, which is the system employed in Cambridge Colleges, was to be the work of the preceptors. They would be constantly in touch with the undergraduates and could soon find out if they were not working. These, however, were only preliminaries to a much more fundamental scheme which Wilson had drawn up after consultation with the Professors at Princeton. On this scheme the Clubs were to be abolished and replaced by a number of Colleges, bearing to the University much the same relation as the Colleges at Oxford and Cambridge bear to those universities.[1] The Colleges were to be places where the undergraduates lived and had their meals. Each College was to have its own Master and staff of preceptors, and was to be an independent unit. Each undergraduate was to be a member of a College and live in it, and the members of a College were to have all their meals together, and to make this possible for the poorer students, these were to be of a very simple character. In American universities there are few,

[1] This system has since been introduced with success into Harvard University.

if any, scholarships for undergraduates, so that these, if they are not supported by their parents, have to earn enough themselves to enable them to stay at the university. A considerable number did this by acting as waiters in summer hotels, tram conductors during rush hours, delivering the morning newspapers and so on. This all takes time and work and those who had to do it were seriously hampered in their studies. To reduce the necessity for this extra work it is very important to keep university expenses down as much as possible, and life in Woodrow Wilson's Colleges would necessarily be austere and a great contrast to that which had prevailed in their Clubs. In Oxford and Cambridge the problem of the poor student is solved by granting scholarships to undergraduates who have shown ability. There are a great number of these scholarships. For example, the Cambridge Colleges in 1934–35 granted from their own funds scholarships amounting in value to about £56,000. In addition to this, scholarships are awarded by some of the great City Companies and by the educational committees of the County Councils. In this way not a few undergraduates receive from scholarships an income sufficient to support them at Cambridge in a style which enables them to join fully in the social life of the place, to meet men of very different means, experience and opinions, to discover that a man may hold views which you regard as pestilential and yet be a very good fellow ; to get, in fine, a wide knowledge of men and manners which will be of great service to them in after life.

There is a very great deal to be said in favour of this method, but it was not at the time possible at Princeton. Feeling in America regarded holding a scholarship as a loss of independence. It admired the grit and determination of the man who refused to accept money when he could earn it some way or another himself. The scholarship system seems contrary to the American spirit. Up to the time when he made these proposals Wilson's success at Princeton had been unprecedented. He was liked by many and admired by all. He had the ear of the Trustees, who appointed a Committee to consider his scheme and report to them. They reported in favour of the scheme, and in June 1907 all but one of the Trustees voted for accepting the Report. It was not long, however, before opposition arose. This came chiefly from the older Princeton men who had been members of the Clubs. This was only natural ; the Clubs were the places where they had made their most intimate friends ; their most vivid and pleasant recollections of Princeton were associated with them. Of their attachment to and interest in Princeton there could be no doubt. They had found the funds for many beautiful buildings, and year after year had made up the deficit between income and expenditure.

There was at this time another question before the University which made the situation still more difficult. Dean West, a man much beloved by many old Princeton men, had started, some time before, the idea of having a residential College at Princeton for graduates who were engaged in study or research. The President favoured this idea, but thought it essential for the unity of the University that this College should be on the campus close to the other University buildings. Dean West was untiring in his efforts to raise funds for this College and a bequest of $500,000 was received on the condition that the College should *not* be on the campus. The President's influence at this time was so great that this bequest was refused.

The fight grew very fierce and, in its later stages, bitter. The President was adamant : he would not make the slightest concession. This attitude is certainly not diplomatic and, moreover, not likely to be successful. The result of the policy all or nothing is much more likely to be nothing than all, and so it was in this case. Another bequest for the graduates' College, this time for $2,000,000, coupled again with the condition that it should not be on the campus, was accepted by the Trustees, who thought that with such an endowment the College, though it might not be in the best place, could yet be made useful to the University. Wilson's scheme came before the Trustees again in October, and whereas six months before only one had voted against it, now only one was for it. The scheme was dead and the Clubs are still flourishing.

This defeat made his position as President intolerable, and he soon after accepted nomination for the Governorship of New Jersey. He was elected Governor of New Jersey in 1911 and in 1913 President of the United States.

This is a very interesting case. If he had been a diplomat he could probably, with the prestige and goodwill he possessed, have been able to carry a scheme which, though not all he had asked for, would have been sufficiently near it to make him think it would be worth a trial, and would have occupied his attention and kept him at Princeton for several years, and he would certainly not have been President during the Great War.

His career was a great tragedy. The last time I ever saw him was at a Levée at Buckingham Palace in 1919, when he was on his way to the Peace Conference. He was standing next to the King, and the guests filed before him and were presented to him. It seemed then possible, and some thought it probable, that he might in a short time be in a position to exert greater influence over the affairs of the world than any man ever had ; yet twelve months after he had lost his influence even in his own country. It was not the failure to secure the Treaty he proposed that was most significant. That Treaty seemed to ignore the fact that there had been a war, that for four years France and England had been suffering without intermission unparalleled losses of men and money, bitter grief and great distress, that France had been invaded and a large part of it laid desolate. Could it be expected that, with these bitter experiences fresh in their minds, France and England could proceed to discuss the Treaty as if they had not occurred, to put them out of their minds and consider only what would be best for Europe in the distant future ? Far more significant, I think, is his failure to bring America into the League of Nations. Some think that he could have done so if he had stayed in America, or if he had brought with him to Europe prominent members of the different parties : there were even some, among them Mr Arthur Balfour, who thought that, though he failed at the first attempt, he might ultimately have succeeded if his health had not broken down, for though he could not manage the Senate, few men have had his power of enlisting the people. Be these things as they may, the question arises, Might it not have been a good thing for the world if Wilson had succeeded, and is it not possible that in the future he may be greatly honoured as one who pointed out the right way though he could not persuade others to follow it ?

One of the most interesting of my experiences was a visit to the Military Academy at West Point, the place where officers are trained for the American Army. I never had much belief in theories of education, but this shattered such as I had. In the first place, the State pays all their expenses and does not allow the students to have the control of any money : if they wanted a postage stamp for a letter they had, I was told, to go to the office and get it from there. After entering West Point they remain there without a break for nearly two years. To mitigate this isolation the Government provides a Guest House at which friends and relations of the students may stay at the week-end, when generally there are dances. The students have to attend a great many classes, but these are interspersed with riding, games and gymnastics, so that they certainly did not show any of the usual symptoms of over-study, but were a remarkably vigorous, well-set-up body of men.

I went to one of their classes ; there was a belt of blackboards round the room ; in front of them, the cadets were standing at attention : the subject was mechanics. The instructor asked me if I would take the class, but naturally I asked to be excused. He then called on one of the cadets and told him to prove the formula giving the time of swing of the simple pendulum. He began, " I am requested to . . . " but he got no further, for the instructor rapped on the table with his gavel and said, " You're nothing of the kind, you are ordered—go on ". He went on and repeated absolutely verbatim the proof given in the textbook ; even the lettering on the figure was the same. This method of learning by rote seems contrary to any reasonable theory of education, but the irony of it is that it produces excellent results. So much so that there is a great demand for West Point men by large employers of industry throughout the country, and there is great difficulty in holding them for the army. The fact is that discipline is the keynote of the system. The students are experts in discipline, a very important thing where large bodies of men are employed.

We returned from West Point to Princeton in time for the official proceedings of the Sesquicentenary. These lasted for three days and were brilliantly successful, the result of hard work for many months, magnificent organisation and the enthusiasm and generosity of Princeton graduates. The celebration was more than that of the Sesquicentenary of the old Princeton, which was not officially the University of Princeton but the College of New Jersey ; it was that of the birth of the new University of Princeton. A large banner with " Ave Vale Collegium Neocaesariense " on one side, and " Ave Salve Universitas Princetoniensis " on the other, hung over one of the arches, and there were innumerable flags showing the Princeton colours, orange and black. The weather was magnificent, and the tulip trees in the campus glowed in the orange of Princeton. It was rumoured that Mr Burbank, a great American gardener who seemed to be able to produce any kind of flower or fruit to order, had been commissioned to produce a chrysanthemum with both the Princeton colours, orange and black, but had not had quite time enough to get it just right.

The proceedings of the first day began with a sermon in the morning from the President of Princeton University, Dr. Patton. In the afternoon there was an address welcoming the delegates from other universities, given by the Rev. Howard Duffren, while President Elliot offered the congratulations of the American universities and learned societies, and I did the same for the European.

The second day began with the recitation of the Academic Ode by Dr. Henry van Dyke representing the Clio Debating Society, while the oration was delivered by

Professor Woodrow Wilson representing the rival debating society, the American Whig Society. Woodrow Wilson's theme was " Princeton in the National Service ". He maintained that by far the most important thing for Princeton to do was to give the kind of education best adapted for making them good citizens. " It should be a place where to learn the truth about the past and hold debates about the affairs of the present." The oration, which was delivered with great dignity and with excellent elocution, roused the audience to the highest pitch of enthusiasm.

In the afternoon there was a football match between Princeton and the University of Virginia. It was my first introduction to American football and I enjoyed it very heartily. I have attempted on page 189 to give some account of the American game.

The day closed with a torchlight procession over a mile long which the President of the United States and his wife, Mrs Cleveland, came to see. A year had been spent in organising it. It was headed by a corps dressed in the uniform of the " Mercer Blues ", the Princeton regiment that fought in the Civil War. Their leader carried the sword General Hugh Mercer had worn at the battle of Princeton, and was followed by representatives of sixty " classes ", beginning with the class of 1839. There were lanterns and torches of almost every colour, but orange, the Princeton colour, overtopped them all, the others looking like specks of colour on an orange ground.

The morning of the third day was spent in conferring fifty-six honorary degrees. On the recipients and the delegates were to be seen the gowns of almost every university and academy, along with some of a type that had never been seen before. It was, I believe, the first time that doctors' robes had been worn in America, and two at least of the American Professors had, with characteristic thirst for improvement, borrowed, from President Gilman of Johns Hopkins, gowns which he had brought home from Europe, where he had received doctors' degrees from several universities. They took these to their wives' dressmaker and told her to select the best features of each and produce something stylish. The result was remarkable : they were the only doctors' gowns I ever saw that had anything that could be called a good fit. These, however, went in at the waist and then billowed out into something very like a skirt in a very coquettish and most unacademical fashion. The President of the United States made a speech after the conferment of the degrees.

On Friday the delegates left Princeton, and many of them attended in the evening a dinner given by the University Club of New York. It was my good fortune to sit next a very celebrated New York physician, and when he heard that I was going to sail for England the next day, he asked if I was a good sailor. I said so far I had not suffered from sea-sickness. He said, " I was just going to say that I was sorry to hear it, but that is not quite what I meant. I meant that I was sorry that I could not be of service to you, for I have a remedy for sea-sickness." I said I should be very glad to know what it was, for one never knew what might happen. " Well," he said, " as soon as you get settled on board have a Manhattan cocktail and then slowly nibble dry biscuits. As soon as you feel it an effort to do this, take another cocktail and then begin again at the biscuits and so on. The theory of it is that you will be all right as long as you can take the biscuits ; the cocktails are to make you want to do this." I have never had occasion to test the remedy but I mention it because it sounds less disagreeable than any I have heard of. My wife and I sailed next day in the *Lucania* and, after an uneventful passage, arrived at Cambridge in time for the October Term.

Princeton University has developed greatly since the Sesquicentenary : it has now large and well-equipped laboratories in all the main branches of experimental and biological science. It has been able to attract Professors of great distinction, some of them, I am proud to say, old pupils of my own. It has been the source of many important discoveries, and has flourishing schools of research in many branches of science. It has lately received a munificent endowment of research from Mr Louis Bamberger and his sister Mrs Fuld, who have made an initial gift of $5,000,000 to enable students who have received the Ph.D. degree or its equivalent to continue their independent training and to carry on research with adequate support, without pressure of numbers or routine, and unharried by the need of obtaining practical results. A school for research in mathematics was established in 1933 with a staff of very distinguished Professors : it has already attracted a considerable number of students, some of them from Trinity College.

Second Visit to America

My next visit to the United States was in the spring of 1903 when I went to give the first course of lectures under the Silliman Foundation. The lectures were on "Electricity and Matter" and were subsequently published as a book with that title by Scribners.

Yale University is at New Haven, the most populous

city in the State of Connecticut. It was often called the
" Elm City " from the number of avenues of elms which
lined the principal streets. These at the time of my visit
were a prominent and very pleasing feature of the town :
shortly afterwards they were for a time destroyed through
the ravages of the elm disease. It was my good fortune
to be, throughout my visit, the guest of the very pleasant
" Graduates' Club ". The Club House was in the University and was frequented by many of its teachers. From
informal talks with them I got a more intimate idea of the
University than I could have done if I had lived outside.
While I was in New Haven, both the town and University were much excited and disturbed by a situation which
had arisen through a strike of the tramcar drivers. To
avoid inconvenience to the public some of the students
had volunteered to act as drivers during the rush hours.
The tramwaymen protested against this and demanded that
the President of the University should forbid it. He
refused to do so and party feeling ran very high, higher
than I have ever known in an English strike. Shots were
fired and, as far as I could see, there was little or no
sympathy between the men and their employers. The
change which has occurred in this respect is very remarkable. On my last visit to America I found all the great
manufacturing companies competing with each other in the
amenities they offered to their employees. The first
things you must see when you visited their works were
the playing-fields, the recreation hall, the club-houses.
They told you of the weekly dances and dramatic performances, and positively gloated over the amount the
company had spent on these. They were evidently bent
on trying to kill socialism by kindness.

Besides the trouble about the strike, there was an out-

SIR J. J. THOMSON : 1902
from the painting by Arthur Hacker in the Cavendish Laboratory

break of typhoid fever in the town and an attack in the newspapers on the water company for not having taken greater precautions to prevent infection spreading from a cottage on the gathering-ground, where there was said to be a case of typhoid. The manager of the water company was in the Club one evening and said that a man had come to him that afternoon and asked him if he would like to have positive proof that the infection had not come that way. The manager said that was just what he wanted ; he said the company would certainly give a liberal reward, but what was the proof ? The man said he had been up to the cottage, and there certainly was a stream which ran near it and then into the reservoir ; that looked bad, but he had followed the stream and found, before it got to the reservoir, a waterfall of some six feet or so. This, he said, proved that the cottage could not have been the source of contagion, for no microbe could possibly have fallen down it without breaking its blooming neck.

Yale University is one of the oldest in America, having been founded in 1701 when it was connected with the Congregationalist Church, just as Princeton was with the Presbyterian and Harvard with the Unitarian. It is now, like them, undenominational.

Most of the science teaching when I was there was given in the Sheffield School of Science, which is at some distance from the other part of the University. The students' Clubs, as at Princeton, play a very important part in the life of the undergraduates, and to be elected to " Skull and Bones " is regarded as the greatest distinction they can attain. I did not, however, hear from the Professors any complaint about its effect on their studies. The mode of election is Spartan. On

Commemoration Day, when many relatives and friends of the undergraduates come to New Haven, at one stage of the proceedings when the undergraduates and the guests are out on the campus, the secretary of the Club appears, mixes with the crowd and taps one man after another on the shoulder. These are the selected candidates and they leave the crowd and go into the Club House. This is all very pleasant for them, but it is a bitter experience for those who hoped and expected to be elected to have their failure announced in so public a fashion and on such an occasion. I was told of one case when a boy who had been thought by most people, as well as by himself, as certain of election, was so affected by his failure that he committed suicide the same day.

The rivalry between Yale and Harvard is quite as keen as that between Cambridge and Oxford and shows itself in more unconventional ways. I saw at the house of one of the Professors a dog who had been trained to pretend to be sick whenever he heard the name Harvard.

There is a distinct difference in " tone " between Yale and Harvard. As might be expected from its proximity to Boston, which was for long the most important literary centre in the States, the atmosphere of Harvard is more pronouncedly literary than that of Yale. President Elliott, who was for many years President of Harvard, was for nearly the whole of that time the leading authority on education in the States, and was a man whose opinion on very many subjects carried great weight in the country. Harvard was thus very much in the public eye on educational and literary subjects and this was not without its effects on the undergraduates.

On the other hand, Yale had in Willard Gibbs one of the very greatest mathematical physicists of his generation.

He was born in New Haven : his father was a professor at Yale. He himself was first a student, then a Tutor, and, from 1871 to his death in 1903, a Professor in that university. I do not know of any case of a more intimate connection between a man and a university. It was long, however, before his university recognised that he was a great man. He had not been a success as a teacher of elementary students. Indeed it is said that there was at one time a movement to replace him. A prophet is, however, not without honour save in his own country, and Clerk Maxwell in 1876 called attention to the vital importance of Gibbs' work. Maxwell was so impressed with it that he constructed with his own hands a model of Gibbs' thermodynamic surface and sent a copy of it to Gibbs. The original is now in the Cavendish Laboratory. In 1901 the Royal Society of London awarded him the Copley Medal, the highest honour it is in their power to bestow : then at last Yale realised how great he was. It should in justice be said that his papers are by no means easy reading and would hardly be intelligible to those who were not experts in the subject. I had myself personal experience of how little his work was known in his own country. When a new University was founded in 1887 the newly elected President came over to Europe to find Professors. He came to Cambridge and asked me if I could tell him of anyone who would make a good Professor of Molecular Physics. I said, " You need not come to England for that; the best man you could get is an American, Willard Gibbs ". " Oh," he said, " you mean Wolcott Gibbs," mentioning a prominent American chemist. "No, I don't," I said, "I mean Willard Gibbs," and I told him something about Gibbs' work. He sat thinking for a minute or two and then said, " I'd like you to give me another name. Willard Gibbs can't be a man of much personal magnetism or I should have heard of him."

Some of the students who attended my lectures were prominent athletes, and they were good enough to show me something of the methods of training those who were to represent Yale in the Inter-University sports. The captain of the football team showed me how they taught their men to tackle. I saw hurdlers do splits on a mattress with one leg stretched out in front and the other behind to train them to get over the hurdles with their legs nearly horizontal. In America they take a great deal more trouble over the technique of their sports than we do in England.

When I arrived in New Haven the spring flowers in the woods not far from the town were at their best. Bluetts, a flower something like a *Tritelia*, were very plentiful and there were great drifts of a red-and-yellow aquilegia, neither of which grow in England. The spring, however, only lasted for a short time and before the end of my stay was followed by a heat wave. This was tolerable in New Haven, as by taking a ride outside a tram-car just before going to bed one could get cooled down and able to sleep without difficulty. It continued, how-ever, in full force at New York when I was on my way home, and a heat wave in New York is one of the most disagreeable things I know. The nights are hotter even than the days and it was impossible to get sound sleep. I was thankful when I got on the boat and was on my way back to England.

I visited Yale again twenty years later, when I went for the opening of a magnificent new chemical laboratory, and found its growth in the interval had been phenomenal.

The number of Professors had more than doubled, many new laboratories had been built and, perhaps the greatest change of all, there had been a very large increase in the buildings of the University by the erection of the Harkness Buildings. These, which cost many millions of dollars, were the gift of the Harkness family and are intended to supply residential quarters for the undergraduates. The buildings form a series of courts modelled after the most famous courts of Oxford or Cambridge Colleges, and the tower is a copy of the tower of St Mary's Church, Shrewsbury. Architecturally I do not think the scheme has proved very successful. The copies of the courts are smaller than the originals. There is, for example, a copy of the Great Court of Trinity College, Cambridge, on a much smaller scale. Now the impressiveness of the Great Court is due not only to the proportion between the height of the building and the side of the square—a great deal depends on the size of the square. If this were reduced the dignity of the Court would suffer. But, apart from architectural considerations, these buildings enable Yale to accommodate a large number of students in the University itself. It introduces to some extent the College system of Oxford and Cambridge to Yale, each of the Courts of Yale functioning like a College at Oxford or Cambridge. One thing, however, had not changed: the new Professor of Physics, Professor Zeleny, like his predecessor, Professor Bumstead, was an old pupil of mine.

The progress at Yale was typical of that made in most of the other universities in America. The State universities who depend upon grants received from the State, were now receiving very much larger ones, while the eastern universities such as Harvard, Yale and Princeton had all received magnificent endowments from generous benefactors. The teaching staffs of the universities had been increased until in the larger ones there were Professors and teachers in almost every branch of intellectual activity. New laboratories had been built which were comparable with those of any country in the world in size and equipment, and new buildings had been added which were pleasing to the eye and well suited for the purpose for which they were intended.

Besides the increase in the number of Professors there had been a greatly needed increase in their salaries. When I first went to America in 1896 these were quite inadequate. Unless a Professor or his wife had private means, it was impossible for them to live in even moderate comfort or for the wife to have any leisure from household duties. Benefactors were willing enough to leave money for buildings which would bear their names, but money for salaries was hardly procurable. I remember on my first visit, when I had unexpectedly to make a speech at a dinner of a society of very wealthy men, I pointed out how hard it was in a country where rent and servants' wages were as high as they are in America, to live on a salary of £500 a year, which was considered the normal salary in those days for a Professor. They said they did not believe that any of them did, for they all married rich wives. Certainly an experience I had seems to show that they were expected to do this. There was a Chair of Physics vacant in an American University, and two or three English physicists had been suggested for it. The President of the university wished someone to see them, and report on the impression they made at an interview. The American Ambassador asked a few physicists, including myself, to meet them at the American Embassy. The Ambassador took no part in the discussion until it came to the question of the salary ; then he said that though the salary was low, many wealthy men lived near to the university. These men had daughters, and Professors held a very good social position in America so that the successful candidate would have no difficulty in procuring a wealthy wife.

I believe that the normal salary of a Professor is now about $6000. Before the depression in the United States came along the universities had difficulty in finding men to fill their junior posts such as Demonstrators or Assistant Lecturers, as men who had done well in their university course had no difficulty in getting employment in industrial concerns with the possibility of getting a very large income if they made good. It must be remembered, however, that the large salaries were given to those who had shown great powers of organisation, not to those who were engaged on the research work connected with the industry. Again, the tenure of these industrial posts was not nearly so secure as that of a Professorship, not only because firms might get into financial difficulties, but also because unless a man was efficient he was dismissed, and that in commercial posts the standard of efficiency to be reached to avoid dismissal was higher than in university ones.

Football as played in America differs very materially from Rugby football in England. In the first place, the costume is different ; the players are much more padded than they ever were over here, and look rather like the advertisements for Michelin tyres. One of the most important differences in the rules of the game is that in America the captain of the team may, during the game, replace one player by another held in reserve. This is allowable not only when the first player has been injured, but also if he gets tired, and even if he is merely not playing well, so that it not infrequently happens that at the end of the game hardly any of the players have been playing from the beginning. If substitutes are to be allowed for injured players, which at first sight seems very plausible, it would seem to be necessary to make the rule as wide as it is in America, otherwise you would have to define what amount of injury qualified for substitution, which would result in continual wrangling. The rule is defended in America as improving the quality of the play, since if a university can supply two good teams, the men in the team which plays first will play harder if they know they will be replaced when they get tired, than would be advisable if they had to last through the whole of the game. Between two universities of the same size the system may have advantages, but between a large university and a small one it obviously gives much greater advantage to the large one. It would increase the difficulty of arranging matches between the universities, and such teams as the Services, Old Merchant Taylors, Edinburgh University and the like. These find it difficult enough to get together a team of fifteen ; it would be vastly more difficult to get the much larger teams required under the American system.

Another difference is that under our system you are not allowed to tackle anyone who has not got the ball, but in America it was not so. What is called interference was allowed. When one of a team has got the ball and is running with it, the forwards on his side arrange themselves in a wedge-like formation in front of him and try to force a way through the opposing team. This led to many fatal accidents and the wedge formation is now forbidden, but the forwards are still allowed to tackle men who are trying to get to the man with the ball.

FOOTBALL IN AMERICA

The applause at a university football match is carefully organised. Students from the universities competing are grouped together under a leader, and when he thinks the appropriate time has come he waves his baton and his chorus give the " College yell."

In a close match the audience is roused in this way to a state of great excitement. I have seen girls jump on their chairs and shout " Kill him ! kill him ! " when one of the team they did not favour was running with the ball.

The rules for the scrum too are different. The ball is not thrown into the scrum by the half-back of the side who has the ball ; it is carried into the scrum by a player on that side, and he puts it down in the scrum, calling out as he does so a series of numbers, say 5–1–7—this means that he is going to push the ball to the player who is called 5 ; 5 must then push to player 1, and 1 to 7, and so on. If they do not get away with the ball and the scrum has to be reformed, the process is repeated, and even a third scrum may be formed ; but if at the end of this the side with the ball has not gained 10 yards, the other side takes the ball and puts it down in the scrum. This is the reason why at the first time the ball is put down, a pole is placed level with it ; this pole is connected with another by a rope 10 feet long, and this is also driven into the ground in the direction of the enemy goal, with the rope stretched. If after three downs the ball is not past the second post, the other side takes it.

There is much more organised strategy in American football than in ours : each player in a match has a number. What that number is nobody but the other members of his side knows. This enables instructions to be given to the team without the other side being any the wiser. Most elaborate precautions are taken to keep the code secret, and I never heard of a case when it had leaked out before a match.

The training of their football teams is very systematic and goes on even when the football season is over. I saw the training in tackling at Yale when I was lecturing there. The teams were lined up in a large barn ; from the ceiling a heavy sack was hung by a rope. The trainer of the team pulled the sack out a considerable distance and then let it go. Just as the rope got nearly vertical he called out a number and the man with that number had to fling himself on the sack and stop it before it got to the place where it had been before the trainer pulled it away. The result of this is that the tackling in the university matches is magnificent. The same may be said of their drop-kicking, which they also practise under all kinds of conditions. The play in American football is not so open as in ours ; in it scrum play is predominant. One does not often see long runs by the three-quarter backs when the ball passes from one man to the other right across the field, and then back again until the enemy's line is crossed. This to my mind is the most thrilling thing in football, or indeed in any game I have ever seen.

Watching football in America is much more expensive than it is in England. The cheapest seat at an Inter-University match is five dollars. In spite of this the attendance is so great that the income of the football club is very large. I have been told that the Princeton Club in 1932 received a million dollars from this source. Football is, however, the only game which brings in a profit. Rowing, baseball, have all to be supported from the money earned by football. The same is true in Cambridge : the finances of the Rugby Club are very prosperous, and they

are liberal in their grants to other clubs; they even
founded a Lectureship in Classics in the University as a
tribute to the Rev. J. Gray, of Queens' College, who was
their President for many years.

CHAPTER VII

Visits to Canada and Berlin

I WAS elected President of the British Association for 1909, when the meeting was to be at Winnipeg. It was just a quarter of a century since the Association had, for the first time in its history, held its meeting outside the British Isles. Then it met at Montreal with Lord Rayleigh, who at that time was Cavendish Professor in Cambridge, as President, while Lord Kelvin was President of Section A. The Association met again in 1897 at Toronto when Sir John Evans was President. It was striking evidence of the immense strides which Canada had made in a quarter of a century that, at the meeting in 1884, Lord Rayleigh paid a visit to Winnipeg, and found that it had not developed beyond the stage when they were ploughing up the streets to get rid of the ruts and make the surface more even. On our arrival we found it a large and prosperous city, with fine buildings set on broad and well-kept streets, and with a population of 200,000.

My wife, brother and son accompanied me on my visit, and in August 1909 we travelled from Liverpool to Quebec by the C.P.R. boat the *Empress of Ireland*. We sailed up the St. Lawrence to Quebec, and arrived there one evening in the late twilight. The approach to Quebec was so beautiful that after all these years I often try to recall it. The river was quite calm, there was no wind, there was not a cloud in the sky, which made a deep violet background for a crescent moon over the Heights of Abraham.

We stayed the night at Quebec and the next morning walked about the town, which was very different from any we were to see afterwards. The population is mainly French-Canadian. French is the prevalent language : the buildings are old, picturesque and the streets tortuous. There is a Roman Catholic university. The inhabitants are mostly Roman Catholics, though many of them have Scotch names. They were at the time of our visit ardent admirers of Sir Wilfred Laurier, the Prime Minister, who was one of themselves, and they had very exalted ideas of his powers. A story current when we were there was that one of them had asked who would succeed King Edward, and when he was told it would be the Prince of Wales, said, " Why, has he such a pull on Laurier as all that ? "

In the afternoon we went to Montreal, and after staying a few hours there, started on our way to Winnipeg. Thanks to the kindness and hospitality of the Canadian Pacific Railway Company the journey was very enjoyable and interesting, and we arrived at Winnipeg without incident. The railway station at Winnipeg is one of the most cosmopolitan places in the world. It is the one on which the streams of immigrants from all countries in eastern Europe converge: Russians, Lithuanians and also Doukhobors, whom I had never heard of before ; they are a most industrious and well-behaved race, but it is part of their religion to walk in procession on one day in the year without any clothes. This makes a difficult problem for the local police to deal with. Very few of these immigrants know any English when they arrive and some of them do not seem to learn quickly. During our stay in Winnipeg the Mayor gave a dinner to some of the officials of the Association, and Sir William White and I set out to walk to his house, which we were told was quite near our hotel. We got lost and had to ask our way, and we had to ask six people before we found one who understood English.

The educational authorities seem to deal with this question very effectively : all the teaching is in English. They have a very large room, and when a child with no English first comes to the school it is not put in any class, but just sent into this room to play with the other children. After about three months of this, most of them pick up enough English to be able to follow the lessons well enough to profit by them. We were told that sometimes forty-two languages and dialects might be heard in the big room.

Professor Rutherford was President of Section A (Physics) ; Professor H. E. Armstrong, of Section B (Chemistry) ; Dr. Smith Woodward, of Section C (Geology) ; Dr. Shipley, of Section D (Zoology) ; Sir Duncan Johnstone, of Section E (Geography) ; Sir William White, of Section F (Engineering) ; Professor J. L. Myres, of Section G (Anthropology) ; Professor Starling, of Section H (Physiology) ; Lieutenant-Colonel Sir David Prain, of Section K (Botany); and the Rev. H. B. Gray (Education). Among others attending the meeting were Professor Larmor ; Professor Poynting, who gave us a paper on " The Pressure of Light " ; Professor Lamb, Professor MacMahon, Professor MacDonald from Aberdeen, Professor Hobson, Professor Lyman and Professor Bumstead from Yale. Among the foreign visitors I may specially mention Dr. E. Goldstein from Berlin, one of the earliest workers on the discharge of electricity through gases, and who, in the course of researches extending over more than fifty years, discovered many very important phenomena. He was a pupil of von Helmholtz and worked at first in the Physical Laboratory at Berlin and later in the Observatory at Potsdam ; his first paper was published in 1874 and was followed in quick succession by others, each containing some interesting discovery : for these he was awarded the Hughes Medal by the Royal Society in 1908. He died in 1930 at the age of eighty. He read a paper to the Association on the phosphorescent spectra given out by some organic compounds when cooled to the temperature of liquid air. Another foreign visitor was Professor H. Münsterberg, Professor of Experimental Psychology at Harvard. He was exceedingly popular with his students ; indeed they were accused of altering the first line of a well-known hymn to " *Ein Münsterberg ist Unser Gott* ". I think the meeting was successful. The various Presidential Addresses were well attended and the discussions well maintained. As was natural in a meeting at the very centre of wheat production in Canada, a lively interest was taken in the prediction made by Sir William Crookes in his Presidential Address to the British Association at Bristol in 1898, that there would, in the not very distant future, not be enough wheat to support the increased population. Actually, since this prediction things have gone the other way, and the production is more than the population can consume, or rather, can afford to buy. It was very interesting to hear how the breeding of a new kind of wheat, which would ripen a fortnight earlier than the kinds known at present, would increase by millions of acres the area on which wheat could be grown.

The evening Discourses were given by Professor Tutton on " The Seven Styles of Crystal Architecture " ; by Professor Herdman on " Our Food from the Water " ; by Professor H. B. Dixon on " The Chemistry of Flame " ;

and by Professor J. H. Poynting on "The Pressure of Light". These, as well as the Presidential Addresses, were given in the Walker Theatre, one of the largest playhouses in the Dominion at that time. The Members of the Association received very generous hospitality, and had the opportunity of learning much about the conditions of life in the newly developed regions in Canada. One thing that struck me very forcibly was the importance of the work done by the Young Men's Christian Association in this, and indeed in all the other towns I visited in Canada. The Association has a very fine club-house, much larger than is usual in this country, and a Committee which is active in getting into touch with young men when they arrive, inviting them to join the Club, giving them opportunities of making friends and of joining in many social functions. This must mean a great deal to a young man, not only when he first comes to the town but as long as he stays there. Another feature which struck me is the great part which the big hotel plays in the social life of the place. Everyone who wishes to meet a friend in the evening seems to think that the best chance of finding him is to go to the lounge of the hotel. Many politicians either entertaining their constituents, or arguing with each other, are to be found there, and business men meet to fix up their deals. If anyone wishes to entertain, he does so at the hotel and not at his own house. This custom seems to have spread to this country. When I was in Dublin recently the lounge in my hotel reminded me very strongly of those I had seen in Canada nearly thirty years ago.

We got an example of the mineral wealth of Canada when we were at Winnipeg, for it was discovered during our visit that the sand used for spreading over the roads contained platinum. Long before this the railroads in the cobalt region in Ontario had unwittingly been metalled with silver ore.

The hospitality of the Canadian people did not cease with the meeting of the Association, for as soon as that was over a large party of us were taken on a most entrancing trip to the Pacific, and then back again to Winnipeg by a different way. The long train by which we travelled was equipped with an observation car, with cooks, waiters, etc., in fact it was an hotel on wheels. Our practice was to travel from one place to another by night, spend the day in seeing the place at which we had arrived, or perhaps make an excursion and stay two days. Wherever we stopped we were entertained by the residents, and I had to make a speech thanking them for their hospitality. This was usually very pleasant and I quite enjoyed it, but at one place they had erected a triumphal arch of bags of wheat. It had been put up hastily, and I noticed, during the Mayor's speech of welcome, that the arch was shaking in a way which seemed to me to indicate a speedy collapse. As we were standing under the arch I thought it best to make the warmth of my thanks very pronounced, so as to compensate for their brevity. I shall not make any attempt to describe in detail the scenery through which we passed. Those who have made the trip will find it hard to believe that it is surpassed by any in the world. At first we travelled for many hundred miles through the prairies, covered with wheat almost ready for the harvest. Each puff of wind made a wave travel over them to the horizon : we might to all appearance have been sailing over a great ocean, instead of running on a railroad through a vast cornfield. The effects at sunset or sunrise were most beautiful and remarkable. After passing over the prairies we reached Banff, and saw the buffaloes which flourish in the immense park reserved for them. Then came the Rockies and snow mountains, with visits to Glacier and Lake Louise. When we were looking down the lake from near the hotel, the other end, which had been quite distinct, suddenly became obscure, although no fogs or clouds had been visible. A few seconds after we heard a great noise and the end of the lake soon got clear; an avalanche had fallen while we had been looking. Some little way before we reached Vancouver, which was our next stop, we ran by the side of the Fraser River : it was so full of salmon that it was quite red. It seemed almost as if they were so thick that an agile very light-weight might have crossed the river using them as stepping-stones. As we approached the Pacific the climate changed and became much more like that in England. The heat was not so fierce nor the air so dry as it had been before. Vancouver we found a very busy and apparently flourishing place. It had a university, parks, golf-links and garden suburbs. There was a land boom on when we were there, and everyone seemed to think he was going to make a fortune by buying, or getting an option on, some piece of land, and then selling it at a higher price. All classes seemed to be infected by the rage, and prices were forced up until some pieces of land sold for more than they would have fetched if they had been in the heart of London.

A few hours' sail from Vancouver is Victoria. This is not a commercial place but one of fine houses, beautiful gardens and well-kept lawns. It seems quite a haven of rest after the bustle and noise of Vancouver.

We returned to Winnipeg by a different route, travelling from Edmonton by the Canadian Northern Line and going through Calgary to Winnipeg. From Winnipeg we went to Toronto to stay with my old pupil and friend

Professor McLennan, who was Professor of Physics in the University, and had succeeded in inducing the Government of Ontario to find the funds for an exceedingly fine laboratory, large, very well designed, and well provided with instruments both for research and lectures. The laboratory was very elaborately organised. For example, in all but the advanced classes there was a file for each lecture in the course, giving the subjects to be dealt with in the lecture and the order in which they should be taken, mentioning, too, the experiments which should be made and the apparatus which should be used. This method has the advantage that if a Lecturer is unable to lecture, a substitute can give the lecture at very short notice. I should be afraid that this method would tend to make the lectures rather dull. Most people, I think, find that their lectures are more easily followed by their pupils if there is some spontaneity about them, if the Lecturer has to find the words as he goes along, and not read them from a manuscript. It was very delightful to find "Mac" going so strong, and taking such a prominent part in the general business of Toronto University. From Toronto we went to Niagara. We had been living in trains for about thirteen days and had spent most nights travelling. I thought I had slept well in the train, but I more than slept the clock round the first night in the hotel at Niagara. I suppose sleep has two dimensions, depth as well as length, and that the sleep one gets in trains does not rest one as much as the same length of sleep in bed. We returned from Niagara Falls to Quebec through Montreal, and visited the University there. We came home from Quebec by the C.P.R. boat the *Empress of Britain*. What struck me very forcibly on my visit to the farming district in Canada was the hard struggle the farmers had for the first few years they were on the land. Their standard of living, until they had made good, was not higher than that of an English agricultural labourer. Another point was the loneliness of their lives in the winter. This has been relieved now by the discovery of wireless, of the telephone and the production of very cheap motor cars; but at the time of our visit it was very acute. A medical officer told my wife that many of his women patients deliberately unpicked their dresses in the winter so that they might find something to do in sewing them up again; without this they were liable to go off their heads.

Visit to Berlin, October 1910

In October 1910 my wife and I went to Berlin to attend the celebration of the centenary of the foundation of Berlin University, where I was to represent Cambridge University. Just as we were about to start, my wife discovered that the hood of my doctor's gown was much too old and shabby to be worn on such an august occasion, so a new one had to be procured. There was no time to undo the parcel, so it was thrust into my portmanteau just as it came. This led to trouble with the custom-house at the frontier, for the hood when folded up is a peculiar-looking object, and the official when he opened the parcel wanted to know what it was. This took some time to explain; then he wanted to know what it cost. Cambridge tailors do not send in the bill with the goods, so that I did not know, and this again took time to settle, and the result was that we very nearly missed the train. However, we arrived in Berlin safely, where we stayed with Professor and Mrs Warburg. Professor Warburg was the

Director of the " Reichsanstalt ", an institution for research in physics which had been founded by Werner Siemens at the suggestion of Helmholtz. It was the first of the laboratories of the type of the National Physical Laboratory at Teddington in England, or the Bureau of Standards at Washington, intended for researches which cannot conveniently be made in laboratories in universities, such as those which have to be extended over long periods, or which require very elaborate or expensive apparatus, or whose interest is mainly in their technical application, or for the comparison of different standards of physical quantities such as the ohm. Von Helmholtz was the first Director of the Reichsanstalt at Berlin and Professor Warburg was his successor.

The Commemoration began on the afternoon of Monday, October 10, with a religious service in the Dom Kirche. The music was magnificent but the building itself very disappointing.

The reception of the delegates on Tuesday was a very striking scene. The delegates in their academic robes of all colours were in the centre of the room, framed by the members of the various student corps in their uniforms, standing against the walls. There were several speeches and the proceedings closed with the singing of *Gaudeamus igitur*. In the afternoon at 3 o'clock there was a banquet to about 600 guests. Bethmann Hollweg (the Chancellor) was in the chair, with a prince on either side. After he had made a short speech, the loving-cup was passed round. He just took a sip and handed it to one of the princes, who treated it in quite another fashion. He threw his head back and drank and drank until it seemed as if he would never stop. At last he did, and then, holding the cup over the table, turned it upside down, and not a drop fell out.

The effect of this feat on the learned assembly was inde-
scribable : they cheered, they shouted, they waved their
napkins ; some of them even stood on their chairs to do so
more effectively. It was for all the world just like a "bump
supper " after the boat-races in Cambridge.

There were several speeches, but these could not be
called " after-dinner speeches ", for there was one between
each course. They were much more serious affairs than
those in England. The speaker left his place and walked
to a pulpit : when he had got into this, he took out a
manuscript from his pocket and read steadily for about
a quarter of an hour. Then he went back to his place at
table and the next course was served. The only German
I can remember who made a speech and did not read
a lecture was the great classical scholar, Professor
Wilamowitz-Möllendorff. Professor Mahaffy of Dublin,
who replied for the British delegates, made a humorous
speech in German and seemed to enjoy making it. One
of my neighbours at dinner said with great surprise,
" Why, he is laughing at his own jokes ".

The result of these speeches was that, though we had
commenced dinner at 3 o'clock, we had not finished it
when we had to leave to attend a performance of *Figaro*
at the Opera at 8, and this, which went on until 11, closed
a very strenuous day.

On the next day there was a lecture in the morning
and a garden party in the afternoon. We also saw a great
number of Zeppelins, which were parked in a large field.
Few of us, I think, had any idea that Germany had so
many.

In the evening the students had a " Commers ", which
seems to have been somewhat uproarious. The supply of
beer which was provided was on the scale of, I think, two

gallons per student, but before 10 o'clock this had all been drunk, and there were hundreds of thirsty students clamouring for beer and not able to get it. There was a disturbance and some damage was done. When the Kaiser heard of this, he sent for some of the ringleaders next morning and rated them as a College tutor might rate an undergraduate who had got into a scrape, only he did it more effectively. He is reported to have said to them : " Why do you drink so much beer ? Why don't you play games like English students do ? If you did you would not be the disgusting-looking objects you are."

We left Berlin on the next day. There was a railway strike on in France, and though we started in the hope that we might be able to reach Calais, we could not do so and had to go to Ostend, and the crossing was very rough.

The working classes in Berlin evidently disliked the English at this time. One morning when my wife was out driving with our host's daughter, she happened to look back and saw some men shaking their fists at the carriage. On asking what was the reason, she was told, " Oh, they have recognised that you are an Englishwoman ". In military circles the Kaiser himself came in for much criticism for not being sufficiently anti-English. The Crown Prince was their hero, who did not fail in this respect. I have never been able to believe that the Kaiser was responsible for the war. Unless the military party had changed greatly in the interval between 1910 and 1914, they wanted war and wanted it badly. Unfortunately they managed to get their way in the end, but I do not believe that the Kaiser was the instigator.

CHAPTER VIII

War Work: Cambridge during the War

THE work I did in connection with the war was mainly in association with the Board of Invention and Research (B.I.R.). This Board was instituted in July 1915 when Mr Arthur Balfour was First Lord, for the purpose of giving the Admiralty expert assistance in organising and encouraging scientific effort in connection with the requirements of the Naval Service.

The functions of the Board were :

(a) To concentrate expert scientific enquiry on certain definite problems whose solution is of importance to the Naval Service.

(b) To encourage research in directions in which it is probable that results of value to the Navy may be obtained by organised scientific effort.

(c) To consider schemes of suggestions put forward by inventors and other members of the general public.

There was a Central Committee of which Lord Fisher was President. The other members were Sir George Beilby, Sir Charles Parsons and myself. Vice-Admiral Sir Richard Peirse was appointed Naval Member of this Committee in July 1916. The Board had its offices in a house in Cockspur Street which Lord Fisher characteristically rechristened Victory House.

The Central Committee was the Governing Body, and its approval was required before any proposal could be adopted. It was assisted and advised by a panel of scientific experts. The original panel was Professor H. B. Baker, Sir William Bragg, Professor Carpenter, Sir William Crookes, Mr Duddell, Professor Frankland, Professor Bertram Hopkinson, Sir Oliver Lodge, Sir William Pope, Lord Rutherford (then Sir Ernest Rutherford), and Mr Gerald Stoney. Sir Richard Threlfall was added to the panel at a later date. The first Secretary was Captain Crease, R.N., who was succeeded by Sir Richard Paget.

The most urgent need of the Admiralty at the time the B.I.R. was instituted was some method of detecting submarines, and means were taken at once to start experiments with this subject. The most obvious method of detecting a submarine is by the sound it makes. It had long been known that sound travels well through water, and various methods had been devised for detecting it. Thus, if a tube is closed at one end with a diaphragm and lowered into water through which sounds are passing, the diaphragm will be thrown into vibration and produce in the air in the tube sound waves of the same pitch as those in the water. These can easily be detected by the ear or by a microphone. The problem of detecting a moving submarine by the sound it makes is a most difficult and complicated one. In the first place, the vessel which is hunting the submarine itself produces noises when it is not at rest, by its engines, its motion through the water, and so on. In bad weather these drown all others. Thus a microphone submerged in the water and carried along by a ship will give tongue even when there are no submarines in the neighbourhood. One thing in our favour was the remarkable power the human ear can acquire in picking out a particular kind of sound even when it is mixed up with others very much louder. An example of this is that workmen in an engineering workshop where there is a din that nearly deafens those who are not used to it, can talk to each other without raising their voices. We found that observers after long practice acquired the power of picking out the sound of a submarine even when mixed up with other sounds, and that better results were obtained by careful and prolonged training of the observers than by increasing the sensitiveness of the instruments. The sound given out by a submarine is not a pure note, but a noise made up of a great number of notes of different pitches. In such a case as this resonance does not help much, and since the character of the note depends upon the proportions between the intensities of the different notes, any resonance effect will destroy the *quality* of the noise of the submarine and make it more difficult to detect. It would be much easier than it is to detect a submarine, if it were a musical instrument and gave out a definite note.

Even if we can identify a noise as due to a submarine we require, if we are to catch it, to know the direction from which it comes. The velocity of sound through water is about 4·3 times that through air. This will be the proportion between the wave-length in water and air respectively. An opaque object placed in the way of a wave will not cast a definite shadow unless the diameter of the object is many times the wave-length. Thus if we determine the direction of the sound in water by placing an object in its way and finding where the shadow is, we have to use much larger objects than would be necessary in air.

The B.I.R. began their attack on the detection of submarines by obtaining from Sir Ernest Rutherford a report on the various methods which had been employed, or suggested, for detecting sounds in water. He reported that the microphone method was by far the most promising. The Board determined to make arrangements on a substantial scale to develop this method, especially on the side of increasing its power of fixing the direction from which the sounds came. They were fortunate in being able to secure the services of Professor, now Sir William Bragg, as the director of this research. It was arranged with the Admiralty that the research should be made at the Naval Experimental Station at Hawkcraig, where Captain Bryan had been making experiments for the Admiralty on sound detection since June 1915. Huts to serve as laboratories were erected by the B.I.R. for Bragg's work, and it paid the salaries of two trained physicists to assist him in his experiments ; a new workshop and a skilled workman were also provided. The plan was that Captain Bryan should continue his work under the direction and at the expense of the Admiralty, while Professor Bragg's work would be under the control and at the expense of the B.I.R. It was intended that these two branches should co-operate and inform each other of the progress they had made. Professor Bragg took charge in May 1916 and made good progress in the development of detectors. Great difficulties were, however, found in getting opportunities for testing these at sea. The Submarine Committee of the B.I.R., after a visit to Hawkcraig in September 1916, reported that conditions were unsatisfactory. One cause of this was that the Navy was then divided into two parties, Fisherites and Anti-Fisherites, and that every scheme associated in any way with Lord Fisher was regarded by the latter party with grave suspicion and dislike. There was so much prejudice of this kind that I believe B.I.R. was said by the Anti-Fisher party to stand for Board of Intrigue and Revenge. Towards the end of 1916 it was decided to transfer the work of the B.I.R. on submarines from Hawkcraig to Harwich, and a laboratory was built at Parkeston Quay. This change improved matters, but it had involved a loss of several months just at the time when a detector had been devised which, as far as could be tested by experiments on a small scale, promised to be of service in detecting submarines. It was essential that it should be tested by fitting the apparatus to some vessels in the Fleet and seeing how it behaved under service conditions. It was only to be expected that under these conditions defects would be detected which would require further experiments at the Experimental Station to overcome. If it had not been for the delays just mentioned, efficient submarine detectors would have been available months earlier than they were and much loss of life prevented.

It would be much easier to detect submarines if the sound they produce had a definite pitch, for then we could, as in the detection of " wireless " waves, make use of the principle of resonance. Practically, we could make them produce such a note at our will if we directed water waves of a definite pitch on to them and listened to the echo. These waves would have to travel from the ship on the look-out for submarines and back again, and unless some contrivance equivalent to that used for searchlights were adopted, the spreading out of the waves would reduce their intensity so much that it would in practice be impossible to detect the echo except at very short distances. Fortunately we can produce the " searchlight " effect much more easily with short waves than with long ones. If, for example, a flat square plate immersed in water is made to vibrate at right angles to its flat surface so quickly that the wave-length of the waves it produces in the water is small compared with the length of a side of the square, the waves will be concentrated along the direction at right angles to the square. The angle of the cone in which they are confined will be proportional to the ratio of wave-length to the side of the square, so that the smaller the wave-length the greater the concentration. Professor Langevin, whom I am proud to be able to say is an old pupil of mine, discovered, when working in France at the detection of submarines, a method by which oscillations could be produced in plates of quartz, so rapid that their wave-lengths were small compared with the size of the plate, and which allowed a great amount of energy to be put into the vibrations, so that the sound they produced was very intense. This method depends on a recondite property of quartz, which had been discovered years before by an investigation made solely with the object of increasing our knowledge of physics without any thought of practical application. There were many other instances in the war of the practical applications of physical phenomena known previously only to students of the higher parts of physics. Indeed we should expect that any part of our knowledge of the properties of matter or of the laws of physics might receive a practical application. One very important American company, noted for its successful applications of physics to practical purposes, is said to instruct the staff of its research department not to trouble about industrial applications, but just discover something and leave it to the staff of another department to find out how to make it pay, and they generally do.

Besides devising physical apparatus to detect submarines

some experiments were made on more unconventional lines. It was suggested to B.I.R. that if sea lions were able to hear sounds made under water they might be used to detect submarines, and thus these might be hunted down by a pack of submarine hounds. Some performing sea lions were hired and experiments were made with them, at first in swimming-baths, afterwards in Bala Lake in North Wales, and finally in the Solent. The method used was to make a sound under water with a " buzzer ", or by hitting a metal plate with a hammer, and to place food at the source of the sound. Thus by following up the sound the animal would always find food, and it was hoped would get to look on it as the call of a dinner-bell. The sea lions were found to hear quite well under water and were able to detect the sound at as great a distance as the detecting instrument we were using at the time. After a short training they learned to associate sound with food and would swim up to its source. In Bala Lake, where the most successful experiments were made, they would some-times come to the food from three miles away. They were, however, very temperamental ; sometimes they would only come from a fraction of this distance. On hot days they were decidedly less efficient than on cold ; their speed too was slower than might have been expected. They took at least forty minutes to travel three miles, so that any but a very slowly moving submarine would get away from them. If the porpoises which often circle round Atlantic liners come to the ship through hearing the noise it makes, and associating it with food thrown out from the vessel, they would make much better detectors than sea lions since their speed is so much greater. In the Solent the sea lions were a failure : there were generally several ships about, and they kept turning aside to go to the one making the greatest noise at the place where they happened to be swimming. Another suggestion which proved worthless was to put food on buoys shaped like the periscope of a submarine, in the hope that, if a periscope did appear above the surface of the water, flocks of gulls would fly to it in the hope of getting food.

The B.I.R. and the similar institutions in France and America kept in touch with each other by liaison officers : at one time the French one in England was the Duc de Broglie, a very eminent French physicist, and the American officer was my old pupil the late Professor Bumstead, then Professor of Physics in Yale University, while Sir Ernest Rutherford, Sir Richard Paget and Captain Bridge visited America and France for the same purpose.

Besides the Committee on Submarines there were Committees on Aeronautics, Naval Construction, Marine Engineering, Internal Combustion Engines, Oil Fuel, Antiaircraft, Noxious Gases and Ordnance and Ammunition.

Part of the work of the B.I.R. was the examination of the suggestions and schemes sent in by inventors, and the general public, for dealing with problems connected with the war. These were so numerous that they required a large staff of clerks and a number of experts from the Patent Office to cope with them. In the first six months after the formation of the B.I.R. we had over five thousand inventions sent in, and the number increased rapidly as the war went on. I should think before it ended the number had increased to well over 100,000 ; of these not more than thirty proved to be of any value. Though very little that was important for the prosecution of the war came out of this cloud of inventions, its political effect was very considerable. Every invention sent in was examined by experts : no one could say that he had sent in an important invention of which no notice was taken. If there had not been the B.I.R., many would have written to the newspapers, and created an impression that the Government were too casual about the war. Each air raid in London was followed by a crop of hundreds of suggestions for capturing the bombarding Zeppelins. Some of these were very naïve. One was to have large balloons moored over London each carrying thick ropes heavily smeared with *bird lime* and flying at a great height. The idea was that the bombers, when they passed over London, would strike against a rope, stick to it and be captured. Another proposal for ending the war was more elaborate. It was to collect a flock of cormorants, feed them on white food, and peg this in horizontal and vertical lines against the walls of the room in which they were kept. This would give the walls the appearance of brick-work, the food representing the mortar. When they had had sufficient training, they were to be liberated as near as possible to Krupp's works at Essen. The cormorants, when they saw the chimneys, would think the mortar was food, peck it away, the chimneys would fall down, and the Germans, not being able to receive arms and munitions from Krupp's, would surrender.

Proposals like these gave no trouble : they were a comic relief in a very serious and harassing drama. There were, however, others equally ridiculous which gave a great deal of trouble. For example, we received an application from an inventor saying that he had devised a method of preventing aeroplanes passing over our lines : for this he asked £7,000,000. He would not say what the nature of the invention was, but said he would do so after we had given an undertaking that we would construct a piece of apparatus after his plans and, if it did what he claimed, give him the £7,000,000. If we had accepted this offer, it would have obliged us to take skilled mechanics, which were very difficult to get, from important work, and go to the expense of constructing a thing which it was highly improbable would be of any use : we therefore turned it down. Then paragraphs began to appear in the newspapers saying that we had rejected a scheme which might end the war, even though the inventor had agreed to let the payment depend on the scheme being successful. This was followed by questions in the House of Commons, and then by "leaders" in influential newspapers, and a very pretty agitation was worked up. So much so that we received a request which could not be refused, that we should reconsider the question. We therefore decided to have an interview with the inventor. He asked if he might bring his advisers with him : to this we consented. Lord Fisher was not able to be present and asked me to receive the deputation. This proved to be a large one. There was the inventor himself, a mechanic, who I think honestly be-lieved in his invention. He brought with him his financial adviser, a solicitor, an auctioneer—I suppose because he expected an auctioneer to be a voluble speaker—and some other oddments whose vocations I do not remember. I told them that it had been decided to reopen the question, but they must understand that we should not think of taking any steps about it unless we had some information with regard to its nature. If the inventor thought the B.I.R. was too large a body to entrust with the secret, then we must try to find someone acceptable to both parties whose verdict we would accept. Then began a very amusing scene. The inventor was full of eagerness to talk about his scheme, while his financial adviser was con-tinually stopping him, saying he must not give away the secret. I asked if the invention would work however high the aeroplanes might fly. He said it did not matter how high they went provided they did not go so high that gravity ceased to act. This was not very hopeful ; and so it went on; if I asked a question as to whether it would act under certain conditions, the inventor always started off by saying it would, and was stopped by his financial adviser or the solicitor. The other members of the deputation said little or nothing. In the end we agreed to try to find an acceptable umpire. This took some weeks, but finally one was found. This was the invention : to surround our lines by a ring of iron poles, each a quarter of a mile high and separated from each other by the same distance : the poles were vertical and each pole was provided with an engine to make it turn about its axis. Fastened to the top of each pole was a steel chain one eighth of a mile long, and at the free end of the chain there was a bag containing a large quantity of dynamite. When the poles were set spinning by the motors the chains would fly out and become nearly horizontal, so that the troops would be protected by a belt of dynamite. This preposterous scheme had been supported by many influential and honourable men, who however knew nothing technically about the nature of the invention which they thought ought to be investigated. The agitation had attracted a good deal of attention and had been an excellent advertisement for the scheme, and if it had been announced that the Government were considering it, it would probably have been possible to get a considerable response to an issue of bonds which were to be repaid at a high premium out of the £7,000,000 which the Government would pay for the invention.

Another proposal which got considerable support

from some influential people came from an inventor who claimed to have produced gold from quicksilver. It is true that quicksilver in a vessel containing gas at a low pressure sometimes gets coated over with a yellow film of some compound of quicksilver when a current of electricity passes through the gas. To some people everything that glitters is gold, and the inventor, who had observed this, thought he was making gold out of quicksilver. This had been brought to the notice of the Government and again an agitation began, urging them to do something. Mr Arthur Balfour came to me and asked me to go into the matter, as questions were sure to be asked in the House, and the Government would be in a stronger position to answer them if they could say that they had already taken steps to have the discovery investigated. Accordingly I suggested several experiments, and said if these were made in my presence I would be prepared to express a definite opinion on the discovery. These, however, were never made, for it was discovered that attempts were being made to induce people to take shares in a syndicate to exploit the discovery by saying that the King was taking a keen interest in it, while as a matter of fact His Majesty had never even heard of it. The Government thought that this would be quite a sufficient answer to any question that might be raised in the House.

Owing to my connection with the B.I.R., I saw a good deal of Lord Fisher between 1915 and 1918. I never came across anyone with such pronounced personality, nor with such extraordinary driving power. His method was that of the mailed fist rather than the gloved hand, and in carrying out his schemes he made many enemies and hurt many people's feelings. When different schemes

came before him he spent very little time in determining
which should be chosen, and in his choice he seemed to
be guided by instinct rather than by reason. When he
had made his choice, his whole energies were thrown into
carrying it into effect. This was a great contrast to the
practice I had been accustomed to in University matters.
In these, much time and energy is spent in discussing what
scheme should be adopted, so much so that one is apt
to be tired of the scheme before it is started, and to be
languid in carrying it out. There can be no question
which is the better method in war-time. Lord Fisher had
foresight and imagination as well as energy. He could see
the potentiality of inventions which in their early stages had
been nothing but failures. He envisaged what service they
might render if the purely mechanical difficulties were
overcome, as there was a good chance they might be by
skill and perseverance, and he did all in his power to
expedite this process. Thus, in spite of great difficulties
he had been trying type after type of submarines, and it
was, I believe, due to him that we had, at the beginning of the war, submarines in our Fleet. The most
dramatic naval event in the war was the destruction of
German warships at the battle of the Falkland Islands
by the fast cruisers *Invincible* and *Inflexible* which he
had introduced into the Navy. He was strongly in
favour of using oil fuel in our ships, and he often talked
of the desirability and possibility of submersible cruisers.
Though he did so much to introduce into the Navy every
possible mechanical contrivance which could make it more
efficient, yet in his view the tactics of this mechanised
fleet should be as full of the spirit of adventure as those
of the old Navy. He had no use for the motto " Safety
First " : the " Nelson touch " was what he always longed

for. I remember very vividly the morning after the battle of Jutland; when I got to the office he was pacing up and down the room more dejected than any man I have ever seen. He kept saying time after time, " They've failed me, they've failed me ! I have spent thirty years of my life in preparing for this day and they've failed me, they've failed me ! " This was the only time I ever knew him to be doubtful about the issue of the war. He used to say : "The Government, the Army and the Navy may make as many mistakes as they please, but we are bound to come out top in the end because we are one of the *lost tribes.*" He had an excellent knowledge of the text of the Bible ; he used to say that the two things he enjoyed most were listening to sermons and dancing. In the middle of the war, when things were going badly, Mr Winston Churchill spoke in the House of Commons in favour of bringing him back to the Navy, a very notable thing to do, as Fisher's retirement from the Navy was due to differences between his views and those of Mr Churchill with regard to naval policy. The next morning as soon as I got to the office he began, " Have you seen what Winston said about me last night ? " I said I had, and was much surprised. " Surprised ! " he said, " I should think you were. There's never been anything like it since Herod made friends with Pontius Pilate."

Even trivial things he did in the grand manner. The notes he wrote to me from the office were on large-quarto paper put without folding into envelopes of the same size. They consisted of perhaps half a dozen lines, and often ended, "Yours till Hell freezes, Fisher". He wrote, at the time I was working with him, his reminiscences, and asked me to read the first draft. It was the most indiscreet and outspoken document I ever read. I hastened to return it as quickly as possible, for if by any accident it had come into possession of anyone connected with the Press the fat would have been in the fire with a vengeance. Some of his reminiscences were published afterwards in *The Times*, but these must have been rewritten or very severely edited, for they were "but as water unto wine" compared with the draft I saw. I may say that the account he gave of his reasons for leaving the Admiralty was practically the same as that also published later in *The Times* by Mr Churchill, the other party concerned in the affair. He told me also of an incident which had occurred when he was made First Sea Lord. He said Mr Asquith, when making up his Cabinet, sent for him and said, "'Sir John, your name has been mentioned to me for the Admiralty and nothing personally would give me greater pleasure, but I am in this difficulty : I am hoping to get Mr McKenna as my Chancellor of the Exchequer, and I am told that you and he are not on speaking terms'. I said, 'Mr Asquith, I don't know why anyone should have told you that ; it is quite untrue. I never refuse to be on speaking terms with anybody ; you lose so many opportunities of saying disagreeable things.'" I had occasionally to go with Lord Fisher to the Treasury to apply to the Chancellor for financial assistance for the B.I.R., and the relations between them were most cordial, nay, even jovial.

Lord Fisher had had some training in science, for when a young man he had been an instructor in the Navy and given lectures on elementary physics : he told me that he always tried to make them as lively and amusing as possible. If his lectures were like his talk they certainly would be amusing, for he could introduce an un-expected word with great effect: *e.g.* he had received a great number of orders and decorations from foreign courts when travelling with King Edward. "You should see me", he said, "when I have got them all on; I look like a blooming Christmas-tree." Again, I once heard him say, "I cannot understand why X, who is a man of first-rate ability and has done good work, has never received any official recognition : some say it is because he has a wife in every port and never goes to bed sober ; but trifles like that won't explain it".

He seemed to require much less sleep than most people. A naval officer who had been with him a great deal said that in his prime he never took more than four hours' sleep, even in his busiest times. I always got on very well with him, and his grandson's name was entered for admission to Trinity College when he was only two years old.

The experience we had at the B.I.R. showed the danger of leaving the investigation of the applications of science until war breaks out, trusting to being able to improvise some makeshift on the spur of the moment. The transition from the laboratory to the workshop or to the ship is one that in most cases takes a long time and much work and expense. Effects which are of trivial importance in the small-scale experiments in the laboratory, may be vital on the large scale necessary for practical utility. Faraday said of his discovery of the phenomenon of electromagnetic induction that it was a babe, and no one could say what it might do when it grew to manhood ; but it took more than thirty years for it to pass from the nursery of the laboratory to the rough-and-tumble of the workshop. Again, electric waves produced in one room of a laboratory could be detected in another ten years before they could be detected at what now seems the insignificant distance of a mile. For this reason the

B.I.R. got the Government to establish a laboratory for research on problems which, like that of the detection of submarines, might be of service to the Navy. The laboratory is at Teddington, near the National Physical Laboratory. The first director was Mr, now Sir, F. E. Smith, who was succeeded by C. S. Wright, an Antarctic explorer, who is an old pupil of mine.

My experience at the B.I.R. brought home to me how intense and widespread was the eagerness of men of science to do something to help to win the war. Many problems came before us on which it was important to get expert opinion from physicists, chemists, engineers and mathematicians. These frequently involved special investigations, experimental or mathematical, and those who undertook them had to put their own work aside. So far from grudging this, they welcomed the opportunity of doing the war work. We had many appeals for work of this kind from those who had not had such work assigned to them. Again, men who were engaged in great industries made arrangements which enabled them to devote a great deal of time to war work.

A conspicuous example was Richard Threlfall, later Sir Richard Threlfall. He had given up his Professorship at Sydney some time before, and had become a member of the firm Albright & Wilson of Oldbury, near Birmingham, the largest manufacturers of phosphorus in the country, and probably knew more than anyone else in England about phosphorus. He applied this knowledge with great success to war purposes and, through him, phosphorus played a considerable part in the war. It was used for making smoke screens behind which a vessel could hide from an enemy ship. His phosphorus bombs, too, proved very useful. He was also the first to suggest the use of helium in place of hydrogen for airships. Helium is not inflammable and does not explode, and so is a complete safeguard against fire. An airship requires, however, a very large quantity of helium, and at that time there were no appreciable supplies in our Empire. He brought this matter before the B.I.R. and at his instigation we got J. C. McLennan, Professor of Physics at the University of Toronto, an old pupil of mine, to analyse the gases which gush out from the ground in some parts of Canada where there are oil wells, and which are used to light some towns, *e.g.* Medicine Hat. Similar gas streams in Texas had been found to be rich in helium. McLennan threw himself into this work with characteristic energy, and examined a large number of the wells in Ontario, Alberta and British Columbia. The best results were given by a well in the Bow River district in Ontario, where there was about 3 parts of helium in 1000 parts of the gas coming out of the earth. This, though much smaller than that for the Texas wells, is much larger than any known in other parts of our Empire.

Threlfall was a chemist and engineer as well as a physicist, so that his services were in continual request for reports on projects submitted to the B.I.R., and there were few meetings of the B.I.R. when we had not something from him before us. Sir George Beilby and Sir Charles Parsons were, like Threlfall, responsible for large concerns and, like him, spared a great deal of time for war work. After the war and quite independently, indeed for some time unaware that the other was doing it, both Parsons and Threlfall made experiments on an engineering scale to see if they could make diamonds from carbon. The great French chemist, Moissan, thought he had done this. He had obtained small particles which, though they did not look like diamonds, did not behave like pieces of graphite, the other form of carbon. The experiments went on for about two years and were very costly. Threlfall, after he had made up his mind what he had got by his method, happened to meet Parsons and said, " Parsons, I don't mind telling you that my diamonds are graphite ". " So are mine ", said Parsons. I believe the general opinion is that Moissan's must have been so too.

In addition to my work at the B.I.R. I was, for the greater part of the war, Chairman of a Government Committee to report on the position of natural science in the educational system of Great Britain. This was very interesting though it involved attendance at a great many meetings. We examined a large number of witnesses and produced a report which was signed by all the members of the Committee. In the report it stated " that it could now be claimed that some science is taught in all schools and a great deal in a great many. This is a great advance and has practically all been made in the last fifty years. It cannot, however, be said that even now science occupies in our system of education a place commensurate with its influence on human thought and on the progress of civilisation." We pointed out in our report that the examination for entrance scholarships to our great public schools tends to entice boys to classics whose strength may lie in other subjects. In the examination for these scholarships much greater weight is given to classics than to any other subject, and a boy must have spent most of his time on classics if he is to have a chance for a scholarship. Thus when he goes to school he is much further advanced in classics than in anything else, and naturally takes it as his main subject. That this is the case is proved by the fact that, of the entrance scholarships to Cambridge Colleges gained by boys from seven great public schools which give school entrance scholarships, for one gained for science six were gained for classics. This disproportion is far greater than the average for all schools, showing that it is not due to the scarcity of scientific ability as compared with classical, but is an artificial one due to the system in force at these schools.

Another matter which came before the Committee was the intense specialisation adopted by some schools in the preparation of boys for entrance scholarships given by Cambridge and Oxford Colleges. We were told of cases where such boys spent by far the greater part of their time, for two years before the examination, in doing the mathematical papers which had been set in previous years. I think, however, that this mainly occurs in small schools where it is rare to have a boy of anything like scholarship standard. When there is one, the master not unnaturally wishes to make the most of him, so as to raise the reputation of the school.

Another question the Committee considered was that of the age for leaving school. This was not then, as it is now, complicated by its connection with the relief of un-employment. The evidence which came before us showed that, though an extra year was unpopular with a large num-ber of parents on the ground that they would lose the wages which their boys might have earned if they had left school, it was much more unpopular with the boys themselves, who for the most part wanted to be their " own masters " as soon as possible, and thought it more manly to be an errand-boy than to be at school. There were, however, many boys, and some of them by no means the least in-telligent, who were fed-up with school, took no interest in their work, and if they stayed longer at school would be only marking-time until they could get away. Some of these were boys whose minds, like those of very many of their elders, do not exert their full powers until they have some concrete problems to deal with. Then they may show more ability than those who have done better at the bookish subjects and got higher up in the school. I feel convinced that the best subjects for developing a boy's intelligence are those in which he is interested, and if he cannot find these in his school work, I think it is better he should leave school and see whether he cannot find them in business, or in the workshop or the mill. I have seen at the Cavendish Laboratory instances of a great increase in intelligence after leaving school and coming to work as laboratory assistants. I have known cases where a boy, who did not seem very promising when he first came to the Laboratory, at the end of the year showed a decided increase in intelligence. This increase went on until he became a very efficient assistant either for research or for lectures, or a very capable foreman in the workshop. It is remarkable that, though the English are not a specially bookish people, there are many who seem to think that it is only in the study of books that intellect is exerted, in spite of Dr. Johnson's dictum that in the study of history all the higher qualities of the mind are quiescent. The great German man of science, von Helmholtz, who, beginning as a medical student, was led by his medical studies to study physiology, by his physiology to study physics, by his physics to study mathematics, and rose to be one of the foremost authorities of his time in all these sciences, declared that he had spent more intellectual effort in getting an instrument that was out of order to work properly, than he had in framing the theory which the instrument was being used to test.

Science teaching in schools has had many difficulties to overcome. Laboratories for physics, chemistry or biology are generally expensive, and may put a severe strain on the resources of those schools which are not in receipt of a Government grant. It is not, however, necessary, perhaps not even desirable, that school laboratories should be equipped with very elaborate and expensive instruments. An enthusiastic teacher may get excellent results with simple and even with improvised apparatus, just as great discoveries have been made by physicists working in their own houses. Indeed, simple apparatus is more intelligible to the student than elaborate instruments where the physics of the experiment may be hidden by the complexity of the instrument. It is necessary, however, that the instrument, though simple, should be designed so as to be capable of giving accurate results.

Another, though less important, point is that while the teaching of classics and mathematics has a long experience and tradition behind it, the teaching of science in schools has no such tradition, and its methods have had to be developed. I think too much importance may be attached to the consideration of method. The personality of the teacher is the most important thing ; a good teacher will soon find the method which in his hands will give the best results, and will do better with this than with one imposed on him from without.

I was, during the greater part of the war, President of the Royal Society, and for the last eight months of it Master of Trinity College. Though the work of the President of the Royal Society was not quite so arduous as it is in normal times, when he is expected as the representative of British science to attend a large number of public dinners which, especially when he is not living in London, take up a good deal of his time, I had with it, the B.I.R. and the Committee on Education, plenty to do, and spent the greater part of my time in London, generally, however, coming back to Cambridge in the evening, as I found its quiet refreshing after the turmoil of that city. One night, however, after a very long day's work, I stayed in London at Garland's Hotel in Suffolk Street. After I had gone to sleep I was awakened by a prodigious noise, which turned out to have been due to a bomb which fell close to the hotel. I soon fell asleep again and slept until breakfast-time. When I came down I found I was the only person who had not spent the night in the basement. The Zeppelin had returned not long after it had dropped the first bomb and dropped several other bombs as it came back, without awakening me.

Cambridge during the War

With the breaking out of the war in August 1914 there began a period lasting for more than four years when everyone had to give up his usual work and turn to something which might help to enable us to win the war. Those undergraduates who were physically fit joined the Army, taking at first commissions in the Regular Army, and later in Kitchener's. The older men helped with the work in Government offices, e.g. the Foreign Office and the Admiralty. Some who had an especially intimate knowledge of some foreign language, or were adepts at acrostics or cryptograms, joined the department which was established for decoding the German wireless messages; others went as masters in schools to free a younger man for service in the Army. Some of those who remained in Cambridge undertook to patrol the streets at night to see that all lights were out, as it was thought very important to make it as difficult as possible for the Zeppelins to locate Cambridge. In this they were very successful, as Cambridge was never bombed in the war, though bombs fell within a few miles. It was so dark at night that people who, like myself, are bad at seeing in the dark, when they went out, were continually bumping into people in the streets. So vigilant were these inspectors, and none more so than the late Dr. McTaggart, the distinguished philosopher, that Trinity College had to give up using white linen table-cloths on the dining-tables in Hall, as it was thought that light might be reflected from these through the lantern at the top of the Hall. The oak tables looked so well without cloths that these have not been resumed.

The cloisters of Trinity College very early in the war were used for a hospital for some of those wounded in the earlier battles. They continued in use until the large hospital which was being installed on the King's and Clare cricket ground was completed. At first, too, some regiments were billeted in one of the courts of the College, but for by far the greater part of the time the College was filled with young men who had already joined the Army, and came to receive an intensive course of instruction, by lectures and by physical training, to qualify them to receive a commission. Some of them had already been at the Front as privates and had shown promise of making efficient officers. They were under the charge of a military staff. Besides their training, the cadets played football and cricket matches, had athletic sports and published an illustrated magazine, *The Blunderbuss*. This will live in history, as it was in it that Professor Housman, who was in residence, first published his well-known poem beginning

As I gird on for fighting
My sword upon my thigh,
I think on old ill fortunes
Of better men than I.

The cadets attended church parade in the Chapel on Sunday mornings, and they adopted Bunyan's hymn—

He who would valiant be
'Gainst all disaster,

one often sung in our Chapel, as a song to sing when they were on the march. At the end of their course they were entertained in the College Hall at a farewell dinner.

Batches of them came to the Lodge on Sunday mornings after Church parade, and I got an opportunity of talking to them. Some of them were remarkably interesting and intelligent and took a great interest in the buildings and pictures in the College. Some who had been at the Front were miners, and I was surprised to find the affection they had for their mine. They said a mine was such a nice place, so warm and comfortable, and they seemed to dislike the trenches even more than the others.

The stay of the cadets at Trinity gave us many pleasant reminiscences and went off without a hitch. This was due in part to the very friendly relations which existed between the military staff and the Fellows of the College, and in an even greater degree to the tact and diplomacy of the late R. V. Laurence, who took on himself the work of the Junior Bursar, who had gone to the Front. Laurence conducted many difficult negotiations with the War Office with such tact and skill that a solution was reached which was satisfactory to both sides. The loss of Trinity men in the war was grievously large : on the panels in the College Chapel the names of more than 600 Trinity men who fell in the war are inscribed, including three of the younger

Fellows of the College. Keith Lucas, F.R.S., killed in an aeroplane accident, was a College Lecturer in natural science, and a man of remarkable ability. He had done important work in physiology ; he excelled also in designing instruments for scientific research, and invented an internal combustion engine on a novel principle. C. E. Stuart was a College Lecturer in classics ; he was killed at the Front a few weeks after his marriage. Geoffrey Tatham was our Junior Bursar and was much beloved in the College. There is a sundial in the College garden in memory of these three friends. Tatham left a legacy to the College which was used to panel the Combination Room now in use, which was made just after the end of the war by throwing together some rooms which had formerly been a part of the Lodge. It is one of the most frequented rooms in the College : in the daytime it is the place where numberless committees have their meetings, and at night it is where the Fellows take wine after Hall. There is an inscription in the room recording the legacy and there could be no place more fitted to keep his memory green.

In this list there are the names of many younger men who had already distinguished themselves in many walks of life, and done enough to show that much might have been expected from them. They one and all have endowed the College with the precious heritage of being able to count among its members so many who have made the supreme sacrifice for their country. They are benefactors of the College, and it is fitting that the list of our benefactors, which is read every year at our annual Commemoration, should conclude with this reference to them :
" Lastly, while we thus enumerate those who have enriched us of their substance, it behoves us also to commemorate

those other benefactors, an unnumbered multitude who by achievements in literature, science, philosophy and the arts, or by patient continuance in well-doing have brought honour to the House and good report, and more especially those six hundred fallen in war whose names are written on these walls. For it is meet that we should have these also in remembrance, celebrating them in our praises and having them in honour at such times as these."

About 16,000 Cambridge men served in the war : of these 2652 were killed, 3460 wounded and 497 reported as missing or prisoners ; 12 obtained the Victoria Cross, 899 the D.S.O. and 5036 were mentioned in dispatches.

Things were at the worst in the academical year 1917–18. Only 281 students matriculated ; the number of men students had fallen to a fraction of the normal value, and since the greater part of the income of the University comes from fees, the financial position was very serious. The Council of the Senate met throughout the war but dealt, with one exception, only with necessary routine business. It would evidently have been very unfair to bring forward any contentious business, involving as it would voting in the Senate House, when the majority of the younger members were away on military service and unable to register their vote. There was, however, one question on which there might be a difference of opinion which it was necessary to settle before the war ended, and that was for how many terms should the men who had been at the Front be required to reside, after their return to Cambridge, to qualify them for a degree. It was evident that it was only fair that they should not be required to reside as long as those who had not been away on service, and that whether they returned to Cambridge or not would depend on the amount of the reduction. The first proposal made by the Council was severely criticised as not being generous enough, and the Council withdrew this proposal in favour of one which allowed those who had been away on service for four or more terms to count four terms as residence in the University, and that they should be excused from passing the Little-Go.

The German victory at St. Quentin in March 1918 gave little hope of a speedy termination of the war, but " the darkest hour is that before the dawn ", and conditions improved very slowly at first but with ever-increasing rapidity, and the Armistice was signed in November 1918. The Government made liberal grants to help those who had served in the war to come back to the University, and these did so in great numbers. In January 1919, 655 students, and between January and June 1552 students, matriculated. These numbers include 400 naval officers who came to Cambridge to complete their scientific studies, which had been interrupted by the war. There were also with us for the Easter Term about 200 American soldier students. Another instance of the rapidity with which the University filled up was that, in June 1919, 105 students passed Part I of the Mathematical Tripos. The Vice-Chancellor, in his address to the University in October 1920, reported that the University was full to overflowing. In June 1919 the University decided, by 161 votes to 15, that Greek should no longer be compulsory for the Little-Go. This ended a controversy which had been smouldering and occasionally bursting out for more than thirty years.

The war had lasted for more than four years, which is a year longer than an undergraduate's stay in College, and we were afraid that there might be no one to hand down the traditional conventions, to restart the clubs and other forms of undergraduate activity. This fear, however, proved baseless: though these students had gone through the grim experiences of the war and were older than the pre-war undergraduates, it was surprising to see how like their ways were in most things. They talked very little about the war; they seemed almost to wish to blot it out of their lives, and to have just the same experiences as those who came before them. There was no breach of continuity, in fact hardly a bump in the crossing from war to post-war times. As the boat race takes place in March and no one had returned until the middle of January there was no time to make preparations, but the cricket match with Oxford, which is not played until July, came off.

The cessation of the war relieved us from much anguish and anxiety and raised great hopes: we thought that, as we had weathered the storm, the rest would be comparatively plain sailing to prosperity greater than the nation had ever had before. These hopes have certainly not been fulfilled. I think, too, there has been a considerable change in the views about war held by not a few of the younger men. In the war there were in the University some conscientious objectors, but not very many; several of these were Quakers, and the greater number objected to war on religious grounds. There were very few whose sincerity could be questioned; indeed it required great moral courage, or exceptional physical cowardice, to face the odium of being a conscientious objector rather than go to the Front. I think in another war the conscientious objector will be a much more serious difficulty than he was in the last: there will be many who would fight to defend their country if it were attacked, but who would not go into another country and attack it. It is, however, difficult in warfare to rely on defence alone for repelling an attack.

My only brother died before the end of the war. He

was the most unselfish of men and devoted most of his free time to the Hugh Oldham Lads' Club in Ancoats, Manchester, of which he was Honorary Secretary for 21 years. This Club is connected with the Manchester Grammar School, and Ancoats was then one of the most slummy districts in Manchester. My brother, who was a bachelor, was engaged all day in business, but used to spend regularly four evenings a week at the Club. He was a great lover of boys and very successful in his dealings with them. His method was very simple ; he never gave lectures or made speeches, but just talked with the boys one by one. Above all, the boys always knew where they could find him if they were in any difficulty and wanted his help. Playing-fields near Manchester could not be got except at very considerable expense, and the Club had to be content with derelict pieces of land on which " soccer " could be played after a fashion, though cricket was impossible. In games which did not require playing-fields the Club was very successful. Open championships in long-distance races, both swimming and running, were won by boys who had been at the Club, where their swimming had been done in municipal baths, and their running by racing through the streets of Manchester at night. As these long races require special powers of endurance, this shows that even the slums of a large town may produce boys of as good physique as any in the country. My brother was very fond of chess, and he managed to impart his fondness for it to a good many of his boys at the Club. A great event in the Club life was the annual camp, when the boys camped out for a week in the country or at the seaside. The preparation for the housing, feeding, entertaining and providing for the many contingencies which were likely to arise, for a camp of five or six hundred boys, was a very

235

formidable business; it involved dealing with number-less details and required heavy work lasting over a long time. My brother had as long and as successful experience of this work as anyone, and was often consulted by Boys' Clubs who proposed starting a camp. He told me what I should not have expected, that the boys enjoyed the camp most when there was a town of considerable size within easy reach; after two or three days in the country they began to miss the bustle and shops and other attractions of the town. If they could go into a town for an hour or two this longing was cured, and they came back eager again for the camp life. He very often had the camp at Prestatyn, a place on the Welsh coast near to Rhyl, a town which acted as a tonic. I heard of another case which also showed this nostalgia for the town in town boys. A man interested in boys' welfare took two London boys for a trip to the Canadian Rockies. When they first saw the snow mountains he said, "Did you ever see anything like this before? Isn't it magnificent?" One boy said, "It may be all very well for them as likes it, but give me London on a Saturday night".

My brother's health broke down in 1914 and he had to give up his work at the Club. He came to live at Cambridge and spent much time in writing letters to boys at the Front who had been at the Club. He also came across some Cambridge boys he was able to befriend. He died on July 11, 1917.

I lost during the war an old friend, Sir W. D. Niven, F.R.S., who had been very kind and helpful to me ever since I was a freshman at Trinity.

Lord Rayleigh, the Chancellor of the University, died at his home, Terling Place, Essex, on June 30, 1919. He had not been in good health for above a year, but was well

enough to deliver his Presidential Address to the Psychical
Society, of which he was President and had been a member
for 22 years, in April 1919. The funeral was at Terling and the
University sent a deputation headed by the Vice-Chancellor.

Lord Rayleigh had been elected Chancellor in succession to the Duke of Devonshire in 1908. He had some
hesitation about standing because, with one exception (the
Marquis of Camden), there had been no Chancellor for
over two hundred years who had not been a Duke at least.
Curiously enough Mr Arthur Balfour, who succeeded him,
felt scruples about allowing his name to be put before the
University for the Chancellorship because no commoner
had ever held that office. Lord Rayleigh was elected
Chancellor without opposition. The installation was on
June 17 ; he came up to Cambridge on the 16th and opened
in the afternoon a new wing of the Cavendish whose
erection was made possible by the donation of £2000 which
he had made to the University when he received the Nobel
Prize. For five years he had come to the Laboratory
almost daily in term time, but he had never before come
to it in a scarlet gown with two esquire bedells before him,
carrying their long silver maces, head downwards. These
maces, though they bow their heads before the Chancellor,
disdain to do so before a mere Vice-Chancellor, and are
then carried with their heads uppermost.

After the installation on the 17th, the new Chancellor
conferred honorary degrees on a number of distinguished
people. It is the custom for the Chancellor himself to
nominate the recipients of the degrees given at his installation. Mr Asquith, the Duke of Northumberland, the
Earl of Halsbury, Admiral Sir John Fisher, Sir H. von
Herkomer, the Hon. C. A. Parsons, Sir G. O. Trevelyan,
Bart., Sir James Ramsay, Bart., Sir W. Crookes, Mr Rud-

yard Kipling, Mr Alfred Marshall, Professor G. D. Liveing were the recipients of degrees. At the luncheon after the conferring of degrees, Sir Andrew Noble announced that some of Lord Rayleigh's friends, not resident in Cambridge, wished to express the gratification of the scientific world at his election by offering to the University a sum sufficient to provide an annual prize to bear his name. The Rayleigh Prize for mathematics was founded with this donation and is awarded at the same time, and by the same Electors, as the Smith's Prizes. Sir H. von Herkomer, to express his appreciation of the degree he had received, presented to the University a portrait by himself of Lord Rayleigh in his Chancellor's robes, which is now in the Fitzwilliam Museum.

Lord Rayleigh told me that the one thing he regretted about his election was that the presence of the Chancellor in Cambridge interfered so much with the ordinary business of the University, and gave so much trouble, that he was afraid he ought not to come to Cambridge as often as he had done before. I said I was sure that if he came on unofficial visits the University would respect his privacy, and that they would wish to make the duties as little burdensome as possible, and would not expect him to come to Cambridge except on important occasions. As far, however, as I can remember, he never came to Cambridge except on official visits. He came and delivered an address for the Darwin Celebration in 1909 ; to confer honorary degrees in 1911, and in 1912 on two occasions, one to confer honorary degrees, and the other to receive the International Congress of Mathematicians ; he came in June 1914 for the opening by Prince Arthur of Connaught of the new physiological laboratory given by the Mercers' Company. This was his last visit to Cambridge as there were no ceremonial functions during the war.

Though Rayleigh held and adorned other offices besides the Chancellorship, his official work was not comparable in importance with his scientific, which was quite outstanding both in magnitude and importance. This is not the place to describe it in detail, but it cannot be passed over, though all I can give is an appreciation of it in the most general terms and an indication of what seem to me to be some of its characteristics. In addition to his standard *Theory of Sound* in two volumes, there are 445 papers in the *Scientific Papers of Lord Rayleigh*, published by the Cambridge University Press. These cover a very wide range : besides papers on pure mathematics, there are some on general mechanics, elastic solids, capillarity, hydrodynamics, sound, thermodynamics, dynamical theory of gases, properties of gases, electricity and magnetism, optics ; practically on every branch of physics known when he began his scientific work. Not a few of these, such as those on the determination of the electrical units and on the discovery of argon, had involved years of work in the laboratory, and others long mathematical calculation. While most collections of scientific papers rest undisturbed on their shelves, and are monuments, rather than parts of one's working library, there are no books I refer to so frequently as Rayleigh's collected papers. His papers deserve the description, which Maxwell applied to those of Ampère, as being " perfect in form and unassailable in accuracy ". The style is very clear, so clear in fact that the reader may not realise how difficult the problem was unless he has attacked it himself before reading the paper. It is also very concise, and somewhat austere ; there are no frills. It must have been innate, for one of his examiners in the Mathematical Tripos said some of his answers to the questions might have been sent to Press without any revision. A quality which can be retained in the turmoil of that examination must be deeply ingrained.

Rayleigh possessed to a very remarkable extent the power of putting his finger on the really vital point of a question. He was therefore able to simplify the solution by taking a case where, though this point had not been affected, everything else which only increased the mathematical difficulties had been stripped off, and the mathematical difficulties reduced so much that a solution was possible. The same thing appeared in his experimental work. In the apparatus the part which really affected the accuracy of the results was sure to be all right and carefully made, while the other parts might be made up of bits of sealing-wax, string and glass tubes. Another quality for which he was remarkable was soundness of judgment on physical questions. In this among all the men I ever met he was only rivalled by Stokes, a man of much the same type of mind. It was not easy to get an opinion from either of them if you consulted them about some new idea in physics ; they would not commit themselves unless it affected some question to which they had given a great deal of thought, but when it did come it was well worth having. I owe a great deal to the talks I have had with them. There was no difficulty or delay in getting an opinion from Lord Kelvin but it was always adverse to the idea. Rayleigh shared Maxwell's and Kelvin's view of the importance of models as an aid to physical discoveries. He says, " There can be no doubt, I think, of the value of such illustrations both as helping the mind to a more vivid conception of what takes place, and to a rough quantitative result, which is often of more value from a physical point of view than the most elaborate mathematical analyses ".

The establishment of the National Physical Laboratory

owes much to the influence he exerted in its favour : he took a very keen interest in its progress, and presided over the executive committee until shortly before his death. He was President of the Advisory Committee on Aeronautics from its institution in 1909 until his death in 1919. During the war his advice was often asked by the various Government Committees engaged on war work, especially on questions which involved difficult points in hydrodynamics or acoustics. In addition to his scientific attainments he was a most agreeable companion, human enough to have a store of good stories which he liked telling. Some of these are given at the end of the excellent biography by his son.

On March 5, 1918, I was admitted to the Mastership of Trinity College. Trinity is the only College in Cambridge whose Master is appointed by the Crown, and the ritual for his admission to the Mastership differs from that in other Colleges, where the Master is elected by the Fellows. On the day of admission all the gates of the College were closed early in the morning, and no one was admitted to the College without special permission. At 12 o'clock I arrived at the Great Gate in academical costume, but wearing my hood " squared ", the way it is worn by the Proctors, which makes it look very different from the normal hood. The gate was locked and I gave three or four hard knocks at the wicket-gate ; this was opened a few inches by the Head Porter (who had been my gyp when I lived in College), who asked who I was and what I wanted. I told him my name, and gave him the Patent of my appointment, and asked him to take it to the Fellows. He took it, and then closed the door again, and I waited outside and afforded amusement to a crowd which had assembled in the street. After a very few minutes the

Great Gate was thrown open, and the Vice-Master (Dr. Jackson), who was at the head of a procession of Fellows that had waited in the Ante-Chapel, came forward, shook me by the hand and presented me to the Fellows. A procession was formed to the Chapel, with the Vice-Master on the right and I on the left, and the other Fellows following in order of seniority. No one had been admitted to the Chapel and the doors were locked. The Patent of Appointment was then read, and I made the declaration which the Statutes of the College require before a Master can be admitted. I was then admitted to the Mastership by the Vice-Master, took my seat in the Master's Stall and the choir sang the *Te Deum*. I was then escorted by the Fellows to the doors of the Lodge. Owing to the war there were hardly any undergraduates in the Court; these were replaced by cadets of the Officers' Training Corps, who turned out in force; there were some hundreds of them living in Trinity at the time. There was a dinner for members of the College in the evening, when the Vice-Master proposed my health and I replied.

SIR J. J. THOMSON : *circa* 1922

CHAPTER IX

Visit to America in 1923

I WAS invited by the Franklin Institute of Philadelphia to give a course of lectures there in the spring of 1923, and for that purpose my daughter and I sailed in the *Majestic* in April of that year. The *Majestic*, which had been before the war a German vessel, was taken over when the war ended ; she was at that time, I believe, the largest liner afloat. We had not particularly fine weather, but there were some very agreeable people aboard, and we enjoyed the voyage. It was not, however, very nautical, for we saw a great many waiters and very few sailors. The incident I remember most vividly was the arrival of the news, a few minutes after the race had ended, that Oxford had won the boat race ; this was not expected and has not occurred again.

On arriving at New York we found the Secretary of the Franklin Institute, Mr R. Owens, an old pupil of mine, waiting to welcome us ; he took us at once to one of the largest and most delightfully situated hotels in the city ; this was the beginning of a boundless and most charming hospitality, which made my visit one of the most delightful episodes in my life. We stayed a few days in New York before going on to Philadelphia. I was very much impressed by the change in New York since my last visit some fifteen years before. There were so many new buildings that it seemed to be a new city and one of great beauty and dignity. The sky-scrapers were very much more plentiful and very much higher than before, and this, to my mind, was all to the good. I have always admired sky-scrapers and think that they are the greatest contribution of our generation to architecture. I remember very well the first of these, the old " flat-iron " building in New York. I once saw it in a mist thick enough to obscure the surrounding buildings, and it looked like the bows of a great vessel coming out of nowhere and bearing down on a doomed city. America must have been fortunate enough to possess great architects to create, and make so effective, this new type of building.

Broadway, too, had become, I will not say more beautiful, but certainly more effective, in its special mission of forcing itself upon one's attention. The luminous advertisements were so numerous that each side of the street seemed to be ablaze. Piccadilly Circus at its best or worst is to Broadway but as moonlight is to sunlight.

Mr Eglin, a Vice-President of the Franklin Institute, whose kindness in thinking of and providing everything that would add to our pleasure, interest or comfort was one of the main reasons why our visit was so pleasant, invited us one night to dine with him at the Ambassadors Restaurant, and took us after dinner to the Ziegfeld Follies, the most famous theatre in New York. We saw there Will Rogers, whose tragic death in an Arctic aeroplane accident occurred only a fortnight before I am writing this. His performance was quite unlike anything I ever saw, before or since. He had been a cowboy, and in his spare time had practised doing tricks with his rope. In one of his turns at the Follies, he stood holding one end of a rope in his hand and making it go into all kinds of curves ; while he was doing this, he jerked out one short sentence after another about some political, social, or indeed any kind of event that was in the papers at the time; most of them were about things I had not heard of, but I found some of the others not only excruciatingly funny, but also very sensible. I should think they were quite likely to have considerable effect on politics. Some of his criticisms were very shrewd, and so humorous that they would stick in people's memories after political speeches or newspaper articles had been forgotten.

The next day I visited Mr Coffin, who was then President of the General Electric Company, at his home in Long Island ; he had a beautiful and very large garden and woodlands in which he took a great interest. It was too early in the year for flowers out of doors, but there were plenty in the greenhouses. For some time an epidemic among chestnut trees had been raging and he had lost more than a thousand trees ; he showed us some of the planks which had been sawn from them, and they were riddled with holes more than a millimetre in diameter. Mr Coffin took me to have lunch at the Country Club near by. " Country clubs " are a great institution in America, and are to be found near most cities, but I never saw one nearly so luxurious as this ; it had a magnificent club-house, a polo ground with stables, many lawn-tennis grounds, a swimming-pool, and a golf-club where the annual subscription must, I should think, be a record. I was told that even to get on the waiting-list you had to subscribe for a substantial amount of the debentures issued by the club ; some of the envious critics declared that the number of strokes taken by some of the members to get round the course was as stupendous as the subscription. Mr Coffin had very wide interests : they included etchings, of which he had a very fine collection.

During my stay in New York I visited some works where

they were operating a new process for moulding sheets of aluminium into many different shapes. It consisted in passing an enormous current of electricity through the sheet for a very minute fraction of a second ; the current was so great that even in this short time the sheet became plastic and could be moulded by pressure. The process was most interesting as a triumph of electrical engineering. The enormous current was got by short circuiting through the plate the terminals bringing the electric light supply into the works. So perfect were the arrangements that there was no sparking when this huge current was broken and not a trace of flicker in the electric lamps in the building.

We spent the week-end after our arrival in New York with our friends in Baltimore. I was glad to find that Johns Hopkins University had had some share of the prosperity which the other universities had enjoyed, and had been able to erect much larger and better equipped buildings in the outskirts of the town. My old friend Professor Ames was now the President of the University. Professor R. W. Wood, the well-known physicist, the maker of many important discoveries and the author of a very valuable textbook on Light, was Professor of Physics. Professor Wood has written books in a lighter vein than the one on Light; his *How to tell the Birds from the Flowers* does not require any previous knowledge of ornithology or botany : it is very amusing and has had a very large circulation.

After a very pleasant week-end I left on the Tuesday morning to go to Philadelphia for my lectures, visiting several universities on my way. I travelled for the rest of my journey in almost regal splendour. My kind friends at the Franklin Institute had placed at my disposal a private train. It had a dining-room, a parlour, two bedrooms, a kitchen, a cook, a waiter, and a conductor in charge of the train. I had to travel about a good deal for the rest of my visit, as I had promised to lecture at several universities. All I had to do was to tell the conductor, before I went to bed, where I had to be the next morning and what I would like for my breakfast. When I awoke, I found I was at my destination and the breakfast was ready. In this way I revisited Harvard, Princeton and Yale, lecturing at each of them and assisting at the opening of a magnificent new chemical laboratory at Yale. I had the great pleasure of meeting many old pupils and old friends, as well as meeting many interesting people for the first time. It was also most satisfactory to find the great strides that had been made in these universities since my last visit ; in each one of them there were many new laboratories—large and well equipped ; very many new Professorships had been founded and many more opportunities for advanced study and research made available. Above all, they seemed to be inspired with a genuine enthusiasm for research. The large number of discoveries in Physics and Chemistry of first-rate importance made in the United States during the last ten years proves that full advantage has been taken of these increased opportunities.

I visited also the research laboratory of the General Electric Company at Schenectady ; visits to this are always enjoyable, as you are sure to find that the distinguished physicists there have something new and interesting to show you. This time they had just succeeded in making a talking film that would work. They took a film of me while I was saying a " few words ". I suppose I must have been one of the first to take a part in a talkie film. The researches of the scientific staff of the General Electric Company, Dr. Langmuir, Dr. Coolidge, Mr Dushman, have led not merely to results of outstanding industrial importance, but also to discoveries, methods of investigation, and instruments, which have greatly advanced our knowledge of physics.

I visited, too, the research laboratories of the Bell Telephone Company, a vast and wealthy Corporation, which has magnificent laboratories and a large staff of able physicists. I saw there some most interesting experiments on the effect of taking out of the human voice sounds of different pitch. They had filters which would take a note of a particular pitch out of any sounds that passed through them ; by using a number of filters absorbing notes of different pitches they could find what effect withdrawal of any particular note had upon the voice. They showed me experiments in which all the notes within considerable range of pitches were taken out of a voice without markedly altering its character, certainly not nearly enough to prevent one recognising the speaker by his voice.

I think that the research laboratory of a single company is more likely to be successful than one under the control of a combination of companies. The pecuniary inducements are greater; the management is likely to be better too, since an individual is better than a committee for work of this kind. Then they can choose a problem in which they are particularly interested, when they know the defects of the existing method, its inefficiency, its uncertainty, its cost and so on, and they want to find a method which is free from them. Take, for example, the tubes for producing X-rays ; these require the production of cathode rays. In the old method these were produced from the gas in the tube. This gas produced great irregularities ; the discharge caused the gas to be absorbed by the walls of the tube, until the pressure was too low to allow the discharge to go through the tube. It was evident that a great improvement would be made if the gas were eliminated. The problem was solved by Coolidge, who got his electrons to produce the cathode rays by heating a tungsten filament to a very high temperature, when, as was known, it emits a copious stream of electrons. A firm which had been interested in the production of the old form would have developed a technique in the manipulation of the tube, in the glass-blowing and the insertion of electrodes, which would make the investigation of different methods for improving the tube much easier for them than for anyone attempting the problem without this experience.

I reached Philadelphia, where my lectures were to be given, on April 9, and found rooms provided for me in the Bellevue Stratford Hotel. Philadelphia is one of the most important cities in America and is a great centre of business and finance. To the physicist it has another attraction which it can never lose ; it was the cradle of American science. It is where, in the middle of the eighteenth century, Franklin made researches which will connect its name for ever with electricity. Benjamin Franklin was one of the most remarkable men that America or any other country has ever produced ; his activities and successes covered as varied a field as those of Francis Bacon himself. He was not born in Philadelphia, the city with which he is always associated, but at Boston, New England, in 1706. When he was seventeen he ran away from Boston, where he had been apprenticed to his brother, who was a printer, and arrived at Philadelphia nearly penniless. However, he soon got employment in a printing office there, and it was not long before he started as a printer on his own account and was very successful. He was almost self-educated, for he left school when he was ten. He took great pains to learn to write English. The method he adopted was to read an article in the *Spectator*, and then, after he had forgotten the words but still remembered the substance, try to rewrite the article. The method was very successful in his case, for he acquired the power of writing singularly clear and vigorous English, and became a remarkably effective pamphleteer. He said himself that his power of being intelligible was one of the main causes of his success. For more than twenty years he published *Poor Richard's Almanac*, which, besides the usual contents of almanacs, contained a collection of snippets of worldly wisdom such as " A penny saved is a penny gained ". The circulation of this increased to 10,000—far, far greater than that of any other publication in the country.

It was through the discovery of the Leyden jar in 1745 that Franklin's attention was turned to electricity. This jar made it possible to collect far larger quantities of electricity than had been obtained before, and thus increased the magnitude of its electrical effects. In the original form of the experiment, a metal rod dipping into water in a glass jar was charged up with electricity from a frictional electrical machine ; the experimenter, holding the jar with one hand, touched the rod with the other, a spark passed to it and he felt a shock. This, if the jar had not been charged up strongly, was only momentary, but with stronger shocks it might take some time before he recovered complete control of his limbs, and very strong shocks were fatal. The " electric shock " aroused popular interest in physics to an extent that was not reached even by the discovery of Röntgen rays. Showmen travelled all over the country with their Leyden jars and electrical machines, giving people shocks for a small fee. At a fête attended by the King of France, one of the entertainments provided for the guests was to watch a file of soldiers, who had formed a chain by holding hands, leap convulsively and simultaneously in the air when the electricity from the jar went through them. Some people did not like these shocks. One Professor, after he had received one, said he would not have another even if he were given the kingdom of France. For this he is rebuked by Priestley in his *History of Electricity*. " Far different were the sentiments of the magnanimous Mr Boze, who with a philosophic heroism worthy of the renowned Empedocles, said that he wished he might die by the electric shock, and that his death might provide an article for the Memoirs of the French Academy of Science."

An account of various experiments made with the Leyden jar, and one of the jars, was sent in 1746 as a present to the Library Company—a club of young men which had been founded by Franklin soon after his arrival in Philadelphia—by Mr Collinson, a friend of Franklin, who lived in London and was interested in science. Franklin, aided by a few friends, at once began to make experiments and, in 1747, wrote a long letter to Collinson, giving an account of his experiment, and the conclusions as to the nature of electricity to which he had been led. This was followed by other letters in 1748 and in 1749. In 1751 these letters were published in London without Franklin's cognizance by Collinson. They contain a very clear statement of his views of the nature of electricity. Electricity is regarded as a fluid whose particles repel each other. Matter when not electrified contains a definite quantity of this fluid ; if it contains more than this quantity it is electrified negatively if the fluid is regarded as constituting negative electricity, positively if it constitutes positive, while if it

contains less than this quantity it is positively electrified in
the first case, negatively in the second. The electrification
of a body is due to the passage of this fluid into or out of
it. Two bodies which have each more, or which have each
less, than the normal amount of the fluid repel each other,
while two bodies, one of which has less and the other more
than the normal amount, attract each other. A similar
view of the nature of electricity had been put forward a
few months before Franklin's letter by Watson, a Fellow
of the Royal Society of London, and published in the
Transactions of that Society. It was, however, Franklin's
letters, followed shortly afterwards by his remarkable
experiments on lightning, which made his theory almost
universally accepted. This was due to the clearness of his
exposition, and to his showing that it gave a simple explanation of the many phenomena known to be associated with
the Leyden jar, and led to the discovery of many new ones.
This work, which together with his experiment on lightning established his position as a physicist of the very first
rank, must have been done in little over two years. He
does not seem to have paid any attention to electricity
before 1745, and his first letter in 1747 contains the gist
of the theory. He says himself that he never was engaged
in any study that so totally engrossed his attention. He
must have had quite remarkable insight and instinct in
electrical matters.

The service which the one fluid theory has rendered
to the science of electricity, by suggesting and co-ordinat-
ing researches, can hardly be overestimated. It is still
used by many of us when working in the laboratory. If
we move a piece of brass and want to know whether that
will increase or decrease the effect we are observing, we do
not fly to the higher mathematics, but use the simple con-

ception of the electric fluid which would tell us as much as we wanted to know in a few seconds. Modern researches have led to the view that electricity is carried from one body to another by electrons, particles whose mass is exceedingly small compared with that of the lightest atom ; that all these electrons are of the same mass and each carries the same charge of negative electricity. A collection of electrons would resemble in many respects Franklin's electric fluid, the idea of which was conceived in the infancy of the science of electricity.

In 1752, Franklin made in Philadelphia the celebrated experiment of getting sparks and shocks by flying a kite carrying a pointed rod during a thunderstorm. It was characteristic of him that he should at once apply this discovery to useful purposes by inventing lightning conductors, long pointed rods reaching beyond the highest point of a building and in metallic communication with the ground. The paper describing these experiments was sent in to the Royal Society, but was not thought worthy of publication in its *Transactions*. The Society made amends for this in 1771 by awarding him the Copley Medal, the highest honour in their power to bestow, electing him to the Fellowship of the Society, and relieving him from the payment of either the entrance fee or the annual subscription. In one of his visits to London in 1772, the Society appointed him a member of a Committee to report on the best way of protecting buildings from lightning. Cavendish was a member of the same Committee, which reported in favour of sharply pointed conductors. One member, Benjamin Wilson, dissented because sharp points attracted the discharge and that, as we were ignorant of how strong it would be, it was better to keep out of its way than trust to getting rid of it safely when we had caught it. He was in favour of blunt ends. The battle between the sharps and the flats was complicated by political considerations after the war with America began. Since the pointed ends had been invented by a rebel, George III ordered these to be removed from the lightning conductors at Kew Palace and replaced by flat ones. He tried to make the President of the Royal Society, Sir John Pringle, support this change. The President replied that the laws of Nature were not changeable at royal pleasure. It was then intimated to him by the King's authority that a President of the Royal Society who held such views ought to resign, which he did.[1]

We know now that the action of lightning conductors depends upon properties of the electric current which were not known in Franklin's time. The passage of an instantaneous outburst of electricity like a flash of lightning along a long metallic rod may, unlike the flow of a steady electric current, depend on other things besides the electrical conductivity of the rod, which was the only thing taken into account in designing lightning conductors until Sir Oliver Lodge called attention to other important considerations in his lectures before the Royal Society of Arts in 1888, and if these are neglected the lightning conductor may be a danger instead of a safeguard. When the electrical currents are rapidly changing, the effect of ordinary metallic conduction becomes insignificant. The important factor is the self-induction of the circuit : this depends upon the length of the circuit, the longer the circuit the greater the self-induction. Thus a long circuit will offer a great resistance to the passage of the current, however great its metallic conductivity may be when the current is rapidly changing its direction. This means that the passage of such

[1] A. H. Smyth, *Life and Writings of Benjamin Franklin*, vol. i. p. 107.

currents produces large electric forces, which may cause electric sparks to fly off from the conductor, and these might produce serious damage. In his visits to England before the American War, Franklin must have frequently met Cavendish : they served on Committees together, and Franklin dined very frequently with the Royal Society Club, and Cavendish practically never missed a dinner. Cavendish too was at this time working on his paper, " An Attempt to explain some of the Principal Phenomena of Electricity by means of an Elastic Fluid ". This fluid is practically the same as Franklin's, whose paper, though it had been published more than twenty years before, is not mentioned by Cavendish. It is difficult to conceive two men more unlike. Cavendish's interest was wholly concentrated on science ; he took no interest whatever in political or social questions. Franklin, on the other hand, was interested in almost everything. Only a small fraction of his time was taken up with science : he played an important part in statesmanship, in politics, in municipal, social and even military affairs ; he was fond of society and had multitudes of friends. Cavendish went out of his way to avoid meeting people and was a misogynist, which Franklin certainly was not. They differed as much in their scientific work. Cavendish's paper on the electric fluid is mainly mathematical. He sets out his results in a series of Propositions, Lemmas and Corollaries : his object is to get results which can be tested by accurate measurements rather than to discover new phenomena. Franklin, on the other hand, never uses a mathematical symbol, but tests his theory by seeing if it will lead him to new discoveries.

When dissensions arose between the mother country and its American colonies, Franklin was sent twice on missions to England to represent the view of the American settlers. He tried with all his might to get a peaceful solution of the dispute : his efforts to do this failed, and he then threw his energies into helping America to get by war what she could not get without it, and in this he succeeded. During his stay in England he got acquainted with many of the most eminent men in this country. He was very popular with them ; was a guest of the Royal Society Club much more frequently than anyone else—in some years he dined as often as nineteen times with the Club. Those who came across him were much impressed by his ability. After he had appeared before a Committee of the House of Commons to represent the views of Pennsylvania on the Stamp Act, Burke said it was like a schoolmaster being examined by a parcel of schoolboys. During the war he represented America for some years at Paris, and repeated the successes he had met with in London. He became, it was said, the best-known man in France. He had an immense correspondence, much of which has been preserved : it includes not only letters from distinguished men of science and from prominent statesmen, but some from charming ladies beginning " très cher Papa ".

The French Academy lost no opportunity of showing how highly they esteemed him. Not only was he made one of its members, but his presence at their meetings was regarded almost as a royal favour. Whilst he was in Paris an Austrian physician, Mesmer, was arousing great excitement and making large sums of money by séances where those present were thrown into very abnormal mental conditions, and some were cured of diseases. Mesmer ascribed these results to animal magnetism ; he supposed that his fingers emitted a magnetic effluvium which was a cure for many diseases. The King appointed a Committee to report on this matter. On this Committee were several of the most eminent physicians and a number of scientific men, including Franklin and Lavoisier. Franklin drew up the report, which, while admitting that some of the phenomena were genuine, rejected altogether the existence of animal magnetism and ascribed the effects to physiological causes. It is clear that they were examples of what would now be called hypnotism, and that Mesmer was a man who possessed the power of hypnotising people to a remarkable degree. He realised that there was money to be made out of it, and employed some of the methods of the charlatan to make as much as possible. His cures may have been quite genuine, for hypnotism was employed in medical practice to an appreciable extent in France towards the end of the last century.

Physics, however, was only one of many branches of science in which Franklin did good work : he wrote on medicine, geology, aeronautics, agriculture, meteorology, political economy, chemistry, hygiene. He always had a keen eye for the practical application of any of his ideas. " What is the use of it ? " was the first thing he thought about. He upbraided himself if he indulged in " wild flights of fancy ", though he confesses to enjoy forming hypotheses, as they indulge his natural indolence : his most important scientific work, by the by, was the formation of a hypothesis. When he was working with the Leyden jar he applied the spark to electrocution, and was able to kill a turkey weighing ten pounds. He made a multitude of inventions of very varied kinds, of which a new kind of clock, bifocal spectacles, and the Pennsylvania fireplace for preventing heat being lost by going up the chimney, are but a few. He does not appear to have sought to make

money out of them, for he says that the fate of a successful inventor is to be exposed to "envy, robbery and
abuse".

Philadelphia owes to Franklin some of its most important institutions. Soon after his first arrival there, when
a boy of seventeen, he started a kind of Mutual Improvement Society with a few friends. This developed into the
American Philosophical Society, the oldest scientific society
in America, and also into the Union Library of Philadelphia, one of the oldest and most interesting libraries in the
country. The University of Philadelphia grew out of a
College he had started, and the largest hospital in Philadelphia owes its inception to him.

I gave five lectures at the Franklin Institute on " The
Electron in Chemistry " : these were subsequently published under this title for the Franklin Institute, by the
J. B. Lippincott Company. I give in the preface of this
book my reasons for choosing this subject : " It has been
customary to divide the study of the properties of matter
into two sciences, physics and chemistry. In the past the
distinction was a real one, owing to our ignorance of the
structure of the atom and the molecule. The region inside the atom or molecule was an unknown territory in
the older physics, which had no explanation to offer as to
why the properties of an atom of one element differed
from those of another. As Chemistry is concerned mainly
with these differences there was a very real division between
the two sciences. In the course of the last quarter of a
century, however, the physicists have penetrated into this
territory and have arrived at a conception of the atom
which indicates the ways in which an atom of one element
may differ from that of another." The electron is the
dominating factor in this question, so that it is important

to interpret chemical questions in terms of the electrons and their arrangement. This is what I attempted to do in the lectures.

The lectures were very well attended and in the audience were some of my old pupils, and many physicists whose work was quite familiar to me, though I had not till then made their acquaintance.

I had a very busy but very delightful time during my stay in Philadelphia. I lectured at the Franklin Institute every afternoon and gave addresses at two universities, Haverford and Swarthmore, which were not far from Philadelphia. I had the honour of receiving an Honorary degree from the University of Pennsylvania, and two medals, the Scott and the Franklin, from the Franklin Institute. Dinners were given in my honour on three evenings, and at the farewell one my old friend Doctor Ames, of Johns Hopkins University, gave a much too flattering account of the work I had done. After my last lecture on the Friday, I joined my daughter at New York, and on Saturday sailed for England in the *Homeric*. We had bad weather all the way home, but neither my daughter nor myself suffered any inconvenience from it.

The Franklin Institute was founded in 1826. At the beginning its activities were confined to applied science, but it soon extended its scope and dealt with physics and chemistry generally and not merely with their applications. It has since its commencement published a monthly journal in which many valuable papers have appeared, and a very useful feature is that some of these are reports on the recent developments in some special branch of physics and chemistry. The Institute had just before my visit received a very handsome bequest, the Bartol Fund, which is to be devoted to research. They have established a

Research Laboratory under the direction of Professor Swan, in which much interesting and valuable work has been done.

While I was in Philadelphia I took the opportunity of visiting the famous Women's University of Bryn Mawr. This is a residential university for both post-graduate and undergraduate women students. There are obvious advantages in having universities for women alone. They can choose the system which is best suited for women, while in co-educational universities the system has to have regard to the needs of the men as well as the women, and a compromise is adopted which is not the best possible either for the women or the men. I was very favourably impressed by what I saw of Bryn Mawr, and wish that we had one of the same type in England. The university has large buildings and attractive grounds. The students gave one the impression that they were thoroughly enjoying their College life, and did not confine their interests to the classroom. My visit to Bryn Mawr, which was, however, only a short one, left on me the impression that, though there are no male students at Bryn Mawr, the spirit of the undergraduate life was much more like that of the male undergraduates at Cambridge than was that of the women at the few co-educational Colleges I visited.

Bryn Mawr was founded in 1880 and opened in 1885 ; at first nearly all the Professors were men, as in those early days of women's education there were no women qualified for these posts. Now, however (1936), there are as many women as men holding Professorships. The success of Bryn Mawr is mainly due to the late Miss Carey Thomas, who was a very capable woman with a very pronounced personality. She used the mailed fist rather than the gloved hand, and domineered over the staff and the governors. It used to be said that Bryn Mawr had turned out more distinguished Professors than any other university in America, and this was pretty nearly true if you take " turned out " as meaning got rid of. Woodrow Wilson himself was one of her victims. She did not, however, always get the best of it. It is said that at a meeting of the Governors, when she was bringing a new scheme before them, one of them ventured at one stage to offer some criticism. Miss Thomas said, " The point you raise, Mr Brown, is merely a side issue, we must keep to the main question ". At a later stage Mr Brown again raised some objection. " Did not I tell you before, Mr Brown, that this is only a side issue." " Well, if you come to that, Miss Thomas, woman herself is only a side issue."

Miss Thomas may have been ruthless and domineering, but her thirty-five years' untiring work was the main cause of the success of Bryn Mawr.

Harvard

Harvard is more closely connected with Cambridge University than any other American university, since it was founded by a Cambridge man.

Some of those who came from England to America in the *Mayflower* and had originally settled in Boston, after a short time made a new settlement at a place a few miles away, which at first they called New Town, and in 1636 they started a school or college, and changed the name of the place from New Town to Cambridge. In 1638 the Rev. Edward Harvard, M.A., Emmanuel College, Cambridge, one of the settlers, bequeathed to the College his library and £600, and in consequence it was called Harvard

College. It was attended at first by the children of the Indians in the neighbourhood, as well as by those of the settlers. From this humble origin, the College has in three hundred years grown from being not only the oldest, but also in some respects the most influential, in the United States. It has many more undergraduates than either Yale or Princeton, the other universities of its type. It has 170 full Professorships (more than twice the number of those at Cambridge, England), covering almost every branch of literature and science. It has, by general acknowledgment, the finest law school in the world. It has fine scientific laboratories ; it has one astronomical observatory in New England and another in South America, where the climate is better for observation ; it has a magnificent library, and its endowments have increased more than ten-thousand-fold. To come to lighter matters : it is the seat of a social club, the Porcellian, which I am told is about the oldest social club in America. The totem of the club is the pig, and its rooms are full of pigs of all sizes and materials : gold, silver, copper, ivory, ebony, glass, and many others.

I am glad to say that there are many links between Harvard and Trinity College. Two of the Professors of Harvard are Trinity men—Professor A. N. Whitehead, F.R.S., a Fellow of Trinity College and formerly a Lecturer in Mathematics, and Professor Nock ; and if there are Trinity men at Harvard, there are also Harvard men at Trinity. Mr Lapsley came from Harvard to be a Fellow and a Lecturer in History in 1904. He is still, I am glad to say, at Trinity, and held for the appointed time the very responsible office of Tutor. Again, Harvard has a scholarship, the Fiske Scholarship, the holder of which has to become an undergraduate at Trinity, and we often have other Harvard men among our undergraduates.

The vitality of the links with the past in this part of America is very attractive. You still find people bearing the names of those prominent in American history centuries ago : these are naturally proud of their ancestry.

> I speak of a town in New England,
> The land of the mist and the cod,
> Where the Lowells speak but to the Cabots
> And the Cabots speak only to God.

But you can do more than find the old names ; here and there are still some, though very few, old houses in spacious grounds which have withstood the waves of city extension surging around them, and stand up as a memorial of times and customs which have long passed away. In such a house lived my old friend and pupil, Theodore Lyman, for long Professor of Physics at Harvard, whose experiments on the spectroscopy of ultra-violet have led to results of fundamental importance, and who, besides being a distinguished physicist, is well known as a hunter of big game, and has devoted many vacations to that pursuit. When in 1932, largely through his exertions, the very fine new physical laboratory at Harvard had been completed, he retired from lecturing and became Emeritus Professor. He still, though his mother is now dead, lives with old-time servants in the house which he has lived in since he was a boy. He is a nephew of Alexander Agassiz, the distinguished naturalist, who was also intimately connected with the Calumet and Hecla copper mines. This connection was eagerly seized upon by the advocates of teaching more science in schools, who said that no doubt Agassiz from his knowledge of geology had been able to see that copper was likely to be found near the site of these mines, though no one before had suspected it, and that by buying up land in the district he had made a large fortune which he owed to his knowledge of geology. His sister, Mrs Lyman, when I was staying with her, once told me, after dinner, how her brother really got connected with these mines, and the true story is not nearly so useful for pointing a moral as the old. She said : " When Alexander was a freshman at Harvard, he got acquainted with a happy-go-lucky kind of undergraduate who was always trying to borrow money from his friends, and he borrowed some from Alexander. After a time the borrower's affairs came to a crisis. A meeting was called, and the assets divided among the creditors : as Alexander was the youngest of these, his turn came last, and the only thing left was a piece of land in a place he had never even heard of, much less seen. He protested vigorously against having to take it, but it was that or nothing, and nobody would buy it from him. This was the land on which the Calumet and Hecla mines were afterwards discovered."

President Elliot introduced during his term of office a very interesting scheme called the " Elective Scheme ", in which the candidates for degrees could, with certain restrictions, choose for themselves out of a long and varied list of subjects those on which they would be examined for their degree. This scheme is in accordance with the principle, which I believe is a sound one, that a student gets a better training by studying a subject which interests him than one which does not. The experiment was not very successful, and now the "elective principle" at Harvard has been modified and restricted. It had been found that the subjects which were chosen by the greatest number of students were the " soft options ", the subjects where the number of marks required to qualify for a degree could be obtained with the minimum amount of work.

The fact is that a large number of students have, to

begin with, no special interest in any branch of learning. This interest has to be aroused by the teacher, and to do it he must be a good lecturer, clear, forcible and interesting. Above all, interesting : such teachers are not easily found. Moreover, if the practice in American universities is the same as it is in Cambridge University, surprisingly little attention is paid in elections to Professorships to the powers of the candidates as lecturers. The evidence as to the soundness and extent of their knowledge, and of the value of their own contributions to knowledge, is very carefully considered, but that about their power of presenting a subject in a clear and attractive way does not receive nearly so much attention. I think myself that this is to be regretted. On my view the most important function of lectures is to arouse the interest of the students rather than to impart information ; to make them so interested that they will get the information for themselves if they are told where to find it. I think the lecturer should give the students something that they cannot easily get from books, as, for example, when his lecture is accompanied by ex-periments, or is on some discovery or idea of his own : this perhaps arouses their interest more than anything else. At present I think many of our students are over-lectured. They spend so much time in going to lectures that they have no time to think about the subject for themselves.

All American universities have received great gifts from their old members : the Harvard graduates are so en-grained with this idea that it is said that a wealthy Harvard man would feel that he had tarnished his reputation if he omitted to leave something to Harvard in his will. They lost what would have been perhaps the greatest benefaction that a university ever received, when a man and his wife, who showed no signs of affluence, called on President

Elliott, and said that they had just lost their son and wished as a memorial to him to do something to help other young men to get a good education, and would be grateful if President Elliot would tell them how to do so. The President said that he was afraid that it was impossible to do anything in that way except at a great expense. This suggestion of impecuniosity nettled the lady, and she said, " Well, President Elliot, what has this university cost, buildings, professors and all ? " The President mentioned a great many million dollars, and the lady said, " Come away, Leland, I think we can do better than that ". They went away and founded the Leland Stanford University, with an endowment of 34 million dollars.

CHAPTER X

Some Trinity Men

W. H. Thompson

WILLIAM HEPWORTH THOMPSON, who had been Master of Trinity since 1866, died in 1886. Before his election to the Mastership he had been a lecturer and tutor of the College, Regius Professor of Greek and Canon of Ely. His lectures as College Lecturer, and also as Regius Professor, on Ancient Philosophy, were exceptionally brilliant and attracted very large classes. His published work, however, was small, and in this respect, and indeed in almost every other, he was a great contrast to his predecessor William Whewell, who was a very voluminous writer, very energetic and somewhat overbearing. Whewell was, however, instrumental in introducing many reforms which have proved very beneficial. He was also a great benefactor to the College and erected at his own cost Whewell's Court, the part of the College to the east of Trinity Street. Though he made no discoveries himself, by writing *The History of the Inductive Sciences*, and *The Philosophy of the Inductive Sciences*, he rendered good service to science. Faraday consulted him about the nomenclature he should use in describing his discoveries, and it is to him that we owe the terms electrode, cathode and anode. He was much more successful as an author than as College Tutor; he seems to have known very little about his pupils. In his time the Scholars of the College were elected by the three tutors. One of these, it was said, always voted for his own pupils ; another was so conscientious that he always voted against them for fear of being unduly biassed; while of Whewell it was said that he was quite impartial, for he did not know who were his pupils. He was extremely punctilious about the behaviour of undergraduates when in his presence. If an undergraduate sat down at one of his evening parties, a servant came up and said, " Undergraduates are not allowed to sit in the presence of the Master ". I can vouch for the truth of this, because I have met a man who was present when this was said to a friend of his who had gone with him to a party at the Lodge. This made him unpopular with the undergraduates, but on the death of his second wife, Lady Affleck, they showed so much sympathy that he was greatly touched, and his relations with them became much more cordial. He was killed by a fall when riding : he was a heavy man and the fall proved fatal. I think he must have been in the habit of falling, for one day I noticed in my tailor's shop a medallion of Dr. Whewell. I asked whether he had been a customer of theirs. " Oh yes, sir, he was the best customer we ever had." I said I did not know he was a dressy man. " Well, sir, he was not what you might call a dressy gentleman, but he was one who took to riding late in life."

There can have been few, if any, periods in the history of the College when it had more undergraduates destined to attain outstanding distinction as men of letters than that between 1829, when Thompson came up as a Freshman, and 1832, when he took his degree. Among his contemporaries were Alfred Tennyson, Arthur Hallam, W. M. Thackeray, Lushington, " wearing all that weight of learning lightly as a flower ", FitzGerald, the translator of

Omar, Monckton-Milnes (Lord Houghton), Spedding (the Baconian scholar), W. H. Brookfield,[1] who became a very popular preacher, and who had a great fund of good stories. Thompson, in a letter which has been published, says that on one occasion his hearers were in such convulsions with laughter that they could not sit in their chairs and had to take refuge on the floor. During his Mastership the statutes of the College were twice revised and greatly changed. New statutes were proposed in 1872 and in 1882. The changes proposed by the College in 1872 never came into force as they were disallowed by the Privy Council. The changes proposed in 1882 were in force until 1926, when they were replaced by statutes made by the Cambridge University Commissioners. The changes in the statutes involved weekly meetings of the whole body of Fellows during term time, over which he had to preside. The meetings about each of these changes lasted over a period of more than a year and a half, and some of them were very long; one, with two short adjournments, went on from 11 A.M. to 12 P.M. This entailed a very heavy strain on the chairman, especially as the Fellows were very much divided on the best course to pursue. Henry Jackson, who attended both sets of

[1] I once had a very curious experience with which both Thackeray and Brookfield were concerned. Not long after I became a Fellow, at the end of a busy term, I went to Brighton for a few days to get freshened up. At dinner on the first evening I sat next a middle-aged lady I had never seen before, and had not the faintest idea who she was. After the usual commonplaces, she said, " Now I want you to tell me who is your favourite heroine in fiction ". One does not expose one's deepest feelings to strangers, so, on the spur of the moment, I said Thackeray's Amelia, and was going on to say that I thought her a very well-behaved and good-natured girl. Fortunately before I could do this, she exclaimed, " I'm so glad to hear you say so, because I'm the original Amelia. I'm Mrs Brookfield and both my husband and I were great friends of Thackeray." I sat next her at dinner during my short stay at Brighton, and she told me many interesting things about Thackeray and his domestic troubles.

College meetings, said the passing of the statutes owed much to Thompson's initiative and resolution.

I never saw Thompson until he was an old man, and he was one of the handsomest old men I ever saw : he had silvery-white hair, sharply cut features and a very dignified presence. There is a portrait of him by Herkomer in the Hall of Trinity College. It somewhat accentuates his severity and he looks rather bored. This, I think, was because he and the artist may not have had many interests in common. Herkomer was, however, very well satisfied with his performance, for he told me that he thought two at least of his pictures would keep his memory green. One was this portrait, and the other the portrait of Miss Grant, a tall lady in white satin. The Master is seated in an armchair with his hands gripping the ends of the arms, and this was the cause of a curious incident. Some time after I became Master, I was showing a distinguished physician the pictures in the Hall, and when we came to this one he said, " That man must have had a stroke ". I had been up at Trinity during ten years of Thompson's Mastership and never heard any rumour of such a thing, and said so. The doctor said he was sure he was right, for no one ever gripped a chair like that who had not had a stroke. On enquiry I found the doctor was right, but the matter had been very carefully hushed up.

Though he had a sharp tongue he had a kind heart, and in spite of his glacial exterior he was very human. I did not find this out until some time after taking my degree. When I was an undergraduate I had often been to breakfast parties, which he and Mrs Thompson gave on Saturday mornings in term time, and had thought him so formidable that I tried to get to the other end of the table near Mrs Thompson, who was a very kind-hearted and friendly lady. She must have found it hard work to make these parties successful. Most of the undergraduates were very shy ; many of them she had never seen or heard of before, and did not know what they were interested in, so she had to turn the conversation on to very neutral topics. There was a tale current in my time, that some years before, when among the scholars there was one named Lamb, who became Sir Horace Lamb and a very famous mathematician, and another named Butcher, who became a very distinguished classic and M.P. for the University, she invited both to breakfast and sat between them. In the course of breakfast she said she supposed she ought to feel very nervous, because it must be dangerous to be between the butcher and the lamb. The story went on to say that as soon as she had said this the Master broke in with, " Personal remarks are in the worst possible taste ". I once asked Sir Horace Lamb, who ought to have known, if there were any truth in this, and he said, " Not a word. The tale was concocted at the time by one of the College wits." I think even the shyest of us were very glad to go to these breakfasts, and looked on an invitation as a mark of distinction.

I discovered how human the Master was when, one dreary afternoon in November, after I had taken my degree, I had to go to him to get his signature to a certificate of residence. After he had signed it he asked me to stay, and began telling me stories, some of them very frank, about past and present Fellows of the College, and went on until the bell for dinner in Hall began to ring. He evidently enjoyed chatting in this way, for he asked me to come again any afternoon when I had an hour or two to spare. One of the stories he told me was about Thackeray, who was in his year. The undergraduates at that time were examined at the end of the May Term, and the places in the examination room were arranged alphabetically. This brought Thackeray and Thompson next each other. In the morning the paper was on Elementary Algebra, and Thompson, for a classical man, was a fair mathematician. At any rate, he was good enough to find something to do on this paper, and wrote sheet after sheet. Thackeray whispered to him, " It would be a great help to me if you would turn your papers round so that I could read them ", and Thompson did so. (No emolument of any kind depended on the result of this examination.) The paper in the afternoon was Greek verse. Thackeray, who fancied himself as a classic, went off at a great pace and finished well before the allotted time. As he was going out he handed a copy of his verses to Thompson and said, " You were very kind to me about the Algebra paper this morning ; if these verses are of any use to you they are quite at your service ". The Master, when he told me the story, said, " I had the curiosity to read them, and there wasn't a line without a gross blunder ". He was remarkably imperturbable. I remember a sermon of his in Chapel, when for once he was ending with what was, for him, quite a fervid exhortation. Just before the end, he lost his place and it took him quite an appreciable time to find it. When he did, he started in exactly the same high tone as he had left off. The effect of this syncopated earnestness was quite startling.

To the outside world he was perhaps best known as the sayer of witty things. When Seeley succeeded Kingsley in the Professorship of Modern History, Thompson, when coming away from his inaugural lecture, said, " I never thought we should have missed Kingsley so much ". He said when he was Canon of Ely, " Ely is a very damp place ; even my sermons won't keep dry there ". He could even descend to puns. When the valet Courvoisier was hanged for murdering his master, he said it was the fulfilment of the prophecy, "Every valley shall be exalted".

The most widely known of his sayings, " We are none of us infallible, not even the youngest ", was made at a meeting of the Fellows for the discussion of proposed changes in the statutes of the College. Some have thought that it was levelled at a particular Fellow, and there have been several claimants for this distinction. I was not then a Fellow and did not hear it, but it was much talked about in the College at the time, and I think most people thought that it was impersonal. It was said at the end of a long discussion in which most of the talking had been done by the younger Fellows, and the Master may have thought that his remark embodied a principle which was pertinent to the occasion.

Dr. Butler

Dr. Thompson, whose health had been failing for some time, died in 1886. The Master of Trinity is appointed by the Crown, whereas with one exception the Masters of other Cambridge Colleges are appointed by the Fellows. Proposals have from time to time been made in Trinity that the College should endeavour to have the statutes altered so that the election to the Mastership should be made by the Fellows as at the other Colleges. These have not met with much support. Though it may not be democratic, there are advantages in the present method. When the election is in the hands of the Fellows, and opinion is nearly equally divided, there may be a keen fight. It is possible, indeed it has happened both at

Oxford and Cambridge in a few cases, that this may leave behind embers of bitterness which may destroy the harmony of the College. This, in 1886, was the first vacancy at Trinity since the new statutes came into force, which permitted the election of a layman. At this time there were only two laymen who were heads of Colleges ; now there are only two clerical heads. There was naturally a good deal of speculation in the College about who would be the new Master. I think that perhaps the majority of the resident Fellows hoped it would be a layman, and the names of Lord Rayleigh and Henry Sidgwick were mentioned.

It was, however, offered to and accepted by Henry Montagu Butler, the Dean of Gloucester, who had had an exceptionally brilliant career. When he was at Cambridge he was regarded as the most brilliant undergraduate in residence. He was Senior Classic, had won a University Scholarship for Classics, had been one of the " Apostles " and was elected to a Fellowship at his first try. In 1889, when he was only twenty-six, he was elected Head Master of Harrow in succession to his father, and filled this post with brilliant success for twenty-six years. He was made Dean of Gloucester by Mr Gladstone in 1885, and Master of Trinity by Lord Salisbury in 1886, and was admitted as Master on December 3, 1886. After dinner that evening his health was proposed by the Vice-Master, Coutts Trotter, an old Harrovian. The Master's reply made a very favourable impression. It showed warm affection for the College, paid tribute to the merits of the late Master and of Dr. Whewell, spoke very modestly about himself and said that the College must not look to him for originality or research. Any gifts he had were of a lighter kind, but such as they were they were

all at the disposal of the College. The task the new
Master had undertaken was no light one. The University
and the College were very different from what they were
when he left Cambridge twenty-seven years before. The
College under the statutes of 1882 was governed as far
as its normal business was concerned by a Council consisting of 8 members elected by the Fellows, and 4 *ex-
officio* members, with the Master as Chairman. He had a
casting vote and on some occasions his vote counted for
two, otherwise he had no powers beyond those of a chairman of a Governing Body. Under the statutes in force
when Dr. Butler left Cambridge, the Master had much
greater powers and the Fellows less, for then membership
of the Governing Body went by seniority and not by
election.

I am afraid that at first Dr. Butler did not find presiding
at the meetings of the Council very pleasant. The members of a body which, like the Council, meets very frequently, easily get into the habit, like the members of a
large family, of saying what they think without taking
much trouble to put it into the most polite forms, and to
one who came in from the outside they might be thought
rude. This did not, however, affect their friendship.
Those who had been squabbling at the Council Meeting
would, at the lunch after the meeting, be the best of friends.
To Dr. Butler, however, who was the most courteous of
men, these outbursts were very distressing. This, I think,
was soon realised by other members of the Council, and
they were careful to carry on their discussions in a way at
which no one could take offence, a practice which has been
the custom up to the present time.

Another thing which caused him some uneasiness was
the reserve, which is often supposed to be characteristic of

Fellows of Trinity, in the expression of their feelings, especially those of approval. It may be that some things go without saying, but it is also true that many of these are the better for being said. If a friend has had a success it is better to send him your congratulations than to leave him to take them as read. Dr. Butler was most scrupulous about this : anyone he knew who received some distinction was sure to have his pleasure enhanced by the receipt from him of a letter such as no one but he could have written, full of charm, grace and kindliness. He was as unrivalled as a letter-writer as he was as a speaker. The number of letters of this kind he wrote must have been very large, for any Trinity undergraduate who won any distinction, either in the College or in the University, was sure to have the pleasure of receiving one. It seemed to him to be so natural to write to his friends to express his appreciation of what they had done that, if he did not receive any such expression, he concluded that what he had done was not appreciated. The Fellows of Trinity were very reticent about expressing their feelings, and the Master felt as if he were surrounded by chilling indifference, and indeed a meeting of the whole body of Fellows is about the most difficult audience to address I know. It is as difficult and depressing as broadcasting a speech or speaking in the dark. I think, however, the Master realised, as the years went on, that he had gained not only the admiration and respect of the Fellows but also their affection.

Dr. Butler's Mastership was characterised by almost boundless hospitality and generosity. In a letter to Dr. Westcott dated May 9, 1887, he writes: " When I accepted my very peculiar post it seemed to me clear that hospitality on a large scale, rightly understood, was one of my plain duties. It seemed to me that the Lodge as time went on ought to bring together the leading members of the University, the Fellows and Scholars, friends from a distance, leaders in good causes, whether here or away. . . . I determined that parties at the Lodge should be very numerous and very various. . . . As some little test of variety I may just say to-day, as on Saturday, I have large parties to meet Sir George Trevelyan, our Honorary Fellow ; on Friday I have a meeting of perhaps eighty or one hundred to work for the Toynbee Hall Settlement ; on the 14th the Roundells, Godley, Sir M. Ridley and Charles Dalrymple come. On the 17th we have a large party with which I think you will sympathise. I hope it will begin a yearly institution. I am inviting the newly elected scholars to meet the Vice-Master, the Tutors and some of the older Fellows. . . . On the 21st the George Hamiltons, Fowell Buxtons and others come. . . . Last Saturday we had forty of the Colonial Delegates." In addition to his own hospitality he increased that of the College, for it was largely due to him that the College instituted in 1889 the "Old Boys" dinner, to which members of the College who had kept their names on the College books were invited in groups to dine and sleep at the College. Efforts are made that as far as possible they shall be given the rooms they had when they lived in College. These dinners have proved a very great success ; the guests welcome the opportunity of meeting old friends, and of keeping in touch with the changes and progress of the College. At these dinners Dr. Butler's speech was one of the greatest attractions, for on occasions like these he was unrivalled as a speaker. He always said the right thing in the right way. He had a marvellous memory and had read very widely. Whatever might be the day on which he had to make a speech, he would always remember some interesting event of which it was the anniversary. He could enliven and drive home the point he wished to make by a happy quotation from some great statesman, orator or poet. The manner in which his speeches were delivered added greatly to their effect. He had a very agreeable voice and until near the end of his life it was easy to hear him. The dignity of his presence and manner were an important factor in the success of his speeches, and never more so than at the " Old Boys " dinner when, " Erect in his scarlet robes at the centre of the High Table, the three great circles of silver plate gleaming on the panelled wall behind him, he seemed worthily to represent before the world the majesty of the College ".

The Master preached several times a term in the College Chapel. His sermons had the felicity in phrasing and wealth of illustration of his speeches. Added to this there was an earnestness and reverence which made them very impressive. They were very simple and practical and seemed to me just the right kind for an audience of undergraduates. He made the sermons in Chapel much more attractive than they were before he became Master. Then, sometimes, we had sermons from a preacher who was so determined to be impartial that, when preaching on some article of Christian belief, he would spend so much time over the arguments that might be advanced against it that he had very little time to give those in its favour. Another preacher was so oblivious of worldly affairs that, when Gilbert and Sullivan's opera H.M.S. Pinafore was at the height of its popularity, he began a sermon by saying with great solemnity, " Never or hardly ever ". The undergraduates, to suppress their emotions, had hurriedly to bury their heads in their hands. The eloquence and

charm of the Master's speeches led to his being besieged
with invitations to preach, to address meetings for the
promotion of all kinds of objects, educational, philan-
thropic and missionary. He was, too, a Governor of six
public schools. This made his name and fame better
known to the educated classes of this country than that
of any other Master of Trinity. There had been Masters
who had been as well or even better known to the class
which haunts the Athenaeum Club, but none whose name
was known to so many men of widely different activities.
He was interested in all these, and while he sympathised
warmly with the efforts of the College to promote learning
and research, since only a small fraction of the undergraduates who passed through the College could aspire
to write great books or make important discoveries, he
felt that another and no less important function of the
College was to train the great majority of the undergraduates so that they should be well fitted to carry on the
work of the country, whether this was in politics, in the
law courts, in the Church, as proconsuls or ambassadors,
in the Civil Service, in teaching, in engineering or commerce. When they obtained distinction in their respective
spheres, he rejoiced as much as he did in the distinction
obtained by those who were working at more academic
subjects. He took a keen interest in games, as was natural
for one who had been captain of the Harrow Eleven and
was the father and grandfather of captains. His eldest
son, E. M. Butler, played twice in the Oxford and Cambridge Cricket Match and twice represented Cambridge
in the Racquets Match. His grandson, Mr Guy Butler,
ran the quarter-mile three times for Cambridge in the
Oxford and Cambridge sports and was never beaten : he
was the best quarter-miler of his time. I never saw a more

magnificent example of physical energy than the way he tore down the straight and forced himself in front in the last few yards. Though Dr. Butler approved and took an interest in games, he thought that at some schools and universities the hero-worship given to the athlete was excessive and mischievous.

On July 2, 1913, his eightieth birthday, an address signed by all the Fellows and written by Professor Housman, was presented to him. It runs : ". . . They remember . . . above all, the years of genial and dignified maturity during which you have presided over Trinity College, your ardent zeal for its common welfare, your considerate kindness towards its individual members, young and old, and the union of charm and authority with which you have represented it within the University and before the world ; they recall the wisdom and tact with which you have fulfilled the duties of your office, and the prompt and graceful eloquence, issuing from rich stores of reading and memory, with which you have adorned it."

Coutts Trotter

Coutts Trotter, who died in 1887 at the age of fifty, had taken, I think, a larger share than anyone else in the great developments in the opportunities for study and research which took place between 1870 and the time of his death. He took honours, though not very high ones, in 1859, in both the Mathematical and Classical Triposes. After taking his degree, he was for two years a curate in Kidderminster : after his election to a Fellowship in 1862 he went to study in Germany, and attended the lectures of Kirchhoff and worked at physiology in Helmholtz's

TRINITY COLLEGE, SHOWING THE MASTER'S LODGE

laboratory. He was elected to a lectureship in Science in Trinity College in 1869, to a tutorship in 1871, and was Vice-Master at the time of his death. He was not, however, particularly successful either as lecturer or tutor, and he did very little original research. His real work was the help and sympathy he gave to all schemes which seemed to him likely to promote the study of science either in the College or in the University. He had great influence in both ; he took a very prominent part in the affairs of the College, while in the University he was a Member of the Council of the Senate continuously from 1874 until his death, and also a member of every syndicate that was concerned in any way with science. I think he must have spent a large fraction of his time in attending the meetings of these bodies. He was a most useful man on a committee : he knew his own mind and could express his ideas clearly and was exceptionally skilful in drafting resolutions. He was invariably courteous, even when, as was sometimes the case, his proposals met with fierce opposition, which was not always very politely expressed. One who worked with him on many committees, said that it was always he who formed the first plan and drafted the final report. His services to science were not limited to his work on committees ; he took a great interest in the planning and erection of the many laboratories which were built in his time. When the plans of the Cavendish Laboratory were under consideration, he went with Clerk Maxwell on visits to many physical laboratories, so that the new laboratory might be brought abreast of recent progress. Michael Foster has testified to the interest he took in the construction of the new Physiological Laboratory and the laboratories for Botany and Zoology, and the value of his suggestions. One very important service he rendered, both to Trinity College and to the University, was the part he took in bringing Michael Foster to Cambridge. The first step in this is generally believed to have been taken by George Eliot and George Henry Lewis, who were friends of W. G. Clark, the well-known Shakespearian scholar, who was a Fellow of Trinity. They asked him if some post could not be found in Trinity for Foster. This suggestion was warmly supported by Trotter, with the result that the College established a new post, Praelector in Physiology. Foster came to Cambridge as Praelector in 1870 and remained in it until he was made Professor of Physiology in the University in 1883. The establishment of this Praelectorship enabled physiology to be studied in the University much sooner than it would have been if it had had to wait until the University was in a position to found a Professorship in this subject. Trinity College did the same thing for biochemistry by appointing Gowland Hopkins to a Praelectorship in this subject in 1910, and for geodesy by appointing Sir Gerald Lenox-Conyngham to one in 1921.

A very fundamental change in the method of election to Fellowships in Trinity College was to a very large extent due to the influence of Trotter. Until 1874 the election was determined solely by the performance of the candidates in a written examination, but in that year the College decided that candidates could gain credit not only by their performance in the examination, but also by the merits of a dissertation containing an account of original research carried out by them, which they were allowed to submit to the Electors. This has proved so successful that the paper work in special subjects has been abandoned, and the awards decided on the merits of the dissertations. The importance Coutts Trotter attached to research is perhaps even more strongly emphasised by the regulations for the Coutts Trotter Studentship, for which he bequeathed to the College a legacy of about £7000, " for a studentship for the promotion of original research in Natural Science, more especially physiology and experimental physics. The studentship is not to be awarded by examination, and in the election more regard is to be paid to the promise of power to carry on original work than to the amount of work already done." Among former Coutts Trotter students are Lord Rutherford, Lord Rayleigh and Professor O. W. Richardson.

In my opinion, research has great educational value and can be made a good test of a man's mental power. I have often observed very striking mental development in students after they have spent a year or two on research : they gain independence of thought, maturity of judgment, increased critical power and self-reliance, in fact they are carried from mental adolescence to manhood. It is essential, however, that when using the dissertation as a test of mental power, other things should be taken into account besides the scientific importance of the results it contains. This may be due to his teacher having suggested to the candidate a problem which led to perhaps unexpectedly interesting results, and to having helped him out of his difficulties almost as soon as he got into them. In such a case the dissertation may not prove more than that the candidate is industrious and a careful experimenter ; it does not prove that he is capable of making discoveries without guidance. I think when once the research has been started, the student should be encouraged to try to overcome his difficulties by his own efforts, and that the assistance given by the teacher should not be more than is necessary to keep him from being disheartened by failure, and to prevent the work getting on lines which cannot lead to success.

Again, the candidates for Fellowships are allowed to send in dissertations on any subject they please, and the Electors to the Fellowship are faced with the difficulty of comparing the merits of dissertations on subjects as varied, say, as Stieljes Integrals, Byzantine Art, Radio-activity, the Political Life of Sir Robert Peel, and the Flora of a Tropical Forest. Again, the dissertations are reported upon by Referees, and to estimate the value of the report it is necessary to know something of the temperament of the Referee. Some Referees are prodigal in their use of superlatives, others very sparing. Reporting on the same paper, one may say, "that this paper is the most important contribution to the subject made in the century"; the other that "it is quite a creditable piece of work". Those who know the Referees know that the difference in their reports is due to the difference in their way of expressing their views, and not to any real difference in the views themselves. If the predictions of all the Referees had been realised, then at the elections I have myself attended we should have added to the roll of our Fellows four Newtons and three Bentleys. On paper our method of electing Fellows seems hopeless, but, like many things in this country, methods which seem hopeless on paper, work fairly well in practice, and it has been so in this case. We have made mistakes, but they have been surprisingly few.

Until the statutes of 1926 came into force, anyone elected to a Fellowship held it, and received its emoluments unconditionally, for six years. He was not even required to reside in College. If he wished to go to the Bar, the Fellowship would enable him to tide over the lean years before he had built up a practice. Some who have held high legal offices have been enabled to do so by a Trinity Fellowship. Some, too, went into the Civil Service, some into politics. The system gave each Fellow an opportunity of taking up the work in which he felt his strength to lie and in which he was most interested. It also brought the College into touch with the work of the nation, and helped the College to fulfil its duty to the nation by supplying for its service able men who without it might not have been available. Under the system which came into force in 1926 the tenure of the Fellowship is only four years, and for at least three of these the candidate, before he receives any emolument, must produce evidence that he has been engaged in research. This practically compels him to take up an academic career. He would be too old at the end of four years to enter any other profession, and if he did he would not receive any emolument from the College.

Dew-Smith

A. G. Dew-Smith, who for many years lived in rooms in College and was a member of the High Table, was a prominent figure in our Society in the eighties. He was not a Fellow, nor did he hold any University appointment, so that by our statutes he had no legal claim to rooms in College : he was granted these because he rendered important assistance to the School of Physiology by relieving Michael Foster, who was a great friend of his, of much financial and administrative business. He was of a type not often found in our Society, familiar with life in London and especially with Club life. Robert Louis Stevenson, who sometimes stayed in Trinity for the week-end on visits to Sidney Colvin, is supposed to have represented him in Attwater in *The Ebb-Tide*. To my mind he was more like Prince Florizel in *The Dynamiters*. He was a man of fine presence and distinguished manners, and, if he had kept the tobacconist shop in Rupert Street, he would have handed a packet of cigarettes over the counter with the air of a monarch presenting the insignia of a Knight of the Garter to one of his subjects.

Dew-Smith, too, was one of the best photographers of his day, and photographed many distinguished people : his portrait of Professor Cayley was a great success. His most important work, however, was starting a workshop in Cambridge for making scientific instruments. It is of the first importance that a laboratory where research is carried on should have a workshop connected with it. Each new piece of research generally requires apparatus which cannot be bought ready-made from the instrument-makers, but has to be made to order. This leads to delay, checks the progress of the research and increases the expense. At first the Physiological Laboratory had no workshop and no funds to equip one, or pay the wages of a skilled mechanic. Dew-Smith, at his own charge, took a small house in St. Tibbs Row not far from the Laboratory, fitted it up as a workshop, and engaged a very skilful mechanic named Pye, while he himself devoted a good deal of his time to the business side of the workshop and its superintendence. Pye himself was a bit of a "character" as well as a good workman ; he held very decided views about most things, including the merits of those for whom he was making apparatus. He expressed these views quite freely, to the great delight and amusement of his master.

The workshop was very successful. It began by

making comparatively simple apparatus. The laboratory, however, soon required much more elaborate pieces. The construction of these required greater knowledge of science and of mechanical engineering than either Dew-Smith or Pye possessed. These were supplied by Mr (later Sir) Horace Darwin. By his aid the magnitude and scope of the work increased until the workshop in Tibbs Row developed into the Cambridge Scientific Instrument Company, which has done much to increase the advance of science by the accuracy of their workmanship, and their enterprise in producing new types of instruments as soon as the progress of science requires them.

Joseph Prior

Joseph Prior was a very prominent figure in the social life of the College for nearly sixty years and had, I think, been the tutor of more undergraduates than anyone in the history of the College. He was Tutor for sixteen years while the normal tenure is not more than ten. His was the longest tenure since the end of the eighteenth century, and though some tutors before that time had a longer one —for example, Thomas Jones was Tutor from 1787 to 1807 —the number of undergraduates in the College was then very much smaller than it is now. Prior had, I should think, about 750 pupils in his sixteen years and Jones about 600 in his twenty.

Prior was a Cambridge man and was educated at the Perse School in that town. He matriculated in 1854 at the age of twenty, became a Scholar in 1856, was twelfth Wrangler in 1858, elected Fellow in 1860, Assistant Tutor in 1861, Tutor from 1870 to 1886.

He was not a profound mathematician or an ardent reformer, but no one perhaps did more to increase the gaiety of the High Table. It is very difficult to describe how this was done. His talk was quite spontaneous, he rattled away and now and then burst out with something quite unexpected and often very funny. It emptied

> From unsuspected ambuscade
> The very Urns of Mirth.

A very characteristic example was once when Mr Oscar Browning was complaining that he did not know what to do with his books, they were growing so fast; Prior suggested to him that he should try reading them. Mr Thornely, in his delightful *Whims and Moods*, ascribes this to Dr. Thompson, Master of Trinity, but I have heard the story told from time to time in Trinity for forty years and it has always been assigned to Prior. The style, too, is Prior's and not the Master's.

He had a very quick wit, which sometimes extricated him from difficulties which he had got into in his lectures on mathematics. On one occasion, when lecturing on statics, he attacked a problem on finding what the tension in a string would be under certain conditions. He began it some time before the end of the hour, but had not finished it by then. He put the blackboard carefully away, brought it out at the beginning of the next lecture and went on with the problem. After some little time he said to the class, " At last we have got to the equation which will give us T, the tension in the string ". When he had worked at it a little longer, the equation he had found turned out to be

$$T = T.$$

This would have disconcerted many lecturers, but Prior

rose to the occasion and said, " Well, gentlemen, this at any rate shows that my arithmetic was correct."

In his rooms in the Old Court, which had been at one time the College library, he had some good pictures, and fine pieces of old furniture and one of Chantrey's busts of Sir Walter Scott. Towards the end of his life he lived in the house in Trumpington Road, on the outskirts of the town, which has since developed into the Evelyn Nursing Home. He still, however, kept on his rooms, but made very little use of them.

He was a good judge of wine and a very useful member of the committee which has to select the wine to be bought by the College. Sampling these is by no means as pleasant as might be expected. The College buys the wine soon after it has been bottled and lays it down to mature. Some of these young wines, especially the champagnes and clarets, were very nasty, and the better the vintage the nastier they were. I remember that some Château Latour 1874 was so astringent that it was undrinkable for many years, but in the end became the finest claret I ever drank.

Prior died at his house on the Trumpington Road in October 1918. He left his estate (subject to a life interest) to the College, unhampered by any conditions. This bequest has proved very useful as it has enabled the College to support some desirable objects which it could not, owing to limitations imposed by the College statutes, have done out of Corporate Revenue.

Henry Jackson

No one was more associated with Trinity College in the minds of many generations of its leading men than Henry

Jackson. If two of these met after leaving Cambridge his name was more likely to crop up than that of any other Don, and he was the one they were most likely to call upon if they came to Cambridge. It was he who made a very important addition to the proceedings connected with our Commemoration of Benefactors, by giving a party which began after the dinner in Hall and the speeches were over. To this, which was held in two large lecture-rooms thrown into one for the occasion, he invited all the guests at the dinner, and in addition to these a large number of undergraduates. The guests at the dinner included some undergraduates, scholars, prizemen and a few others asked for special reasons, but it was not possible to find room in the Hall for more than a fraction of those we should like to have had with us. Jackson went round the College a day or two before Commemoration leaving cards of invitation to his party on very many undergraduates who had not been invited to the dinner. At the party there was smoking, whist, speeches, songs and whisky. He generally managed to get speeches out of some of the distinguished guests, and I have heard there Cabinet Ministers singing comic songs. There were clay pipes on the tables, and Jackson flitted about the room with a cigar-box in his hand. The party lasted until the small hours of the morning, and sometimes, I believe, he took the few who were left at 2 or 3 o'clock over to his rooms, and kept the proceedings up for another hour or two.

Jackson came to live in College in 1890, when his wife, who was in bad health, had been advised to live in a milder climate than that of Cambridge. His rooms were in Nevile's Court on the North Side, and on the staircase nearest the Library. They were the ones I had just vacated on my marriage in 1890. Here in a sense he kept open house, for he never sported the oak when he was in his rooms. It was his custom to invite those who went to the Combination Room to take wine after dinner, to adjourn to his rooms to smoke and talk ; and very interesting the talk often was. He was an academic Dr. Johnson, quite as emphatic, though perhaps not so epigrammatic, as the " Great Lexicographer ". His hospitality was not confined to the older members of the College, and he welcomed the Bachelors of Arts and the undergraduates, and did much to establish friendly relations between the reading undergraduates and the older members of the College.

His lectures on ancient philosophy were an outstanding feature in the teaching given in the College. After 1871, and until 1882, all candidates for the Classical Tripos were expected to show some knowledge of ancient philosophy. In this period Jackson's lectures were attended by some seventy or eighty students. Two large lecture-rooms were, for these as for his party, thrown into one, and he lectured with his back against the niche which divided the two rooms. His lectures were attended not only by Trinity men, but by the great majority of the classical students in the University. When in 1882 the Classical Tripos was altered by the addition of a second part, in which candidates could specialise in literature and criticism, philosophy, history, archaeology or philology, candidates could get a First Class by obtaining distinction in any one of these subjects. As philosophy was no longer compulsory, the number attending his lectures naturally went down, but he still attracted the most able classics in the University. His influence was shown by the fact that the number of Trinity men who obtained distinction in philosophy in Part II was, in the ten years following the change, greater than the number who obtained distinction in all the other subjects put together. He was a great teacher ; though he produced no *magnum opus* himself, he trained pupils who did, and these will probably be his most permanent memorial. He gave unstinted assistance to his pupils and his friends in the preparation of their books. Dr. Parry, in his Life of Jackson, gives a list of twenty-seven volumes which were dedicated to Jackson by pupils and friends, with warm acknowledgments of the great assistance he had given them.

He took a great interest in University and College politics, and was a member of the Council of the Senate of the University, and also of the Council of Trinity College, for many years. He had, soon after getting his Fellowship, been most active in attempts at reform, spent a good deal of time in support of abolition of religious tests for Fellowships, the abolition of compulsory Greek, the admission of women to the University and many other reforms. He took a prominent part in debates, both in the Senate House and at College meetings ; he was vigorous and outspoken in these, and sometimes ruffled the temper of those who did not agree with him.

He succeeded Jebb as Regius Professor of Greek in 1906, but continuous bad health and the war interfered seriously with his professorial work. He received the Order of Merit in 1908.

Perhaps the best expression of the feelings with which he was regarded in Trinity are to be found in an address, written by Professor Housman, and presented along with a copy of Porson's tobacco-jar, by the Master and Fellows to him on his eightieth birthday. " In Trinity, in Cambridge, in the whole academic world and far beyond it,

you have earned a name on the lips of men and a place in
their hearts to which few or none in the present or the
past can make pretension. And this eminence you owe
not only or chiefly to the fame of your learning and the
influence of your teaching, nor even to that abounding
and proverbial hospitality which for many a long year has
made your rooms the hearthstone of the Society, and a
guest-house in Cambridge for pilgrims from the ends of
the earth, but to the broad and true humanity of your
nature, endearing you alike to old and young, responsive
to all varieties of character or pursuit, and remote from
nothing that concerns mankind."

Henry Sidgwick

Henry Sidgwick was a very outstanding member of
the College from 1855, when he came up as a Freshman,
until his death in 1900. As an undergraduate he had a very
distinguished career ; he won the Craven Scholarship for
Classics (the blue riband of University scholarships) in his
second year. In 1859 he was Senior Classic, first Chancellor's Medallist and 33rd Wrangler in the Mathematical
Tripos. He was an " Apostle " [1] and President of the
Union. He was elected to a Fellowship at his first try in
1859, and appointed Assistant Tutor in the same year. At
first he lectured on Classics, but after a short time, at the
request of the College, he lectured on Moral Sciences. In
1869 he resigned his Fellowship and Assistant Tutorship, as
his religious opinions had changed since he had signed the

[1] For an account of the society called the Apostles, see *Henry Sidgwick :
A Memoir*, by Arthur Sidgwick and Eleanor Mildred Sidgwick. Macmillan, 1906.

declaration that he believed the doctrines of the Church of England, on his admission to a Fellowship ten years before. As he did not do so now, he thought it was his duty to resign his Fellowship. This action hastened the abolition of religious tests in the College, for in the following year the motion " That the Master and Seniors take such action as may be necessary in order to repeal all religious restrictions on the election and conditions of tenure of fellowships at present contained in the Statutes ", was passed by the requisite two-thirds majority of the Fellows. The College did all in its power to retain him at Trinity : they elected him to a Lectureship in Moral and Political Science and also to an Honorary Fellowship. After the passing of new Statutes in 1882 he was re-elected to an ordinary Fellowship. He was one of the most brilliant talkers of his time. Lord Bryce said his talk " was like the sparkling of a brook whose ripples seem to give out sunshine ".[1] I should think he was, in the opinion of most people, the most brilliant in Cambridge, and Leslie Stephen brackets him with the very eminent mathematician H. T. S. Smith, who was a Professor of that subject at Oxford, and whose epigrams for many years delighted many people. One of them was about an editor of the scientific journal, *Nature*, who was sometimes accused of being cocksure about many things : it was, " X fails to recognise the difference between the Author and Editor of *Nature* ".

Sidgwick had a slight stutter which, whether by accident or design, became much more pronounced just before the point he was about to make ; this brought the point out, as it were, with a bang and made it much more effective. He often took part in discussions at meetings of the Fellows

[1] *Henry Sidgwick*, p. 319.

when suggested changes came under consideration. His speeches were a very enjoyable intellectual treat, but they did not, I think, have much effect on the division. He was sometimes accused of sitting on the fence, but it was rather that he kept vaulting over it from one side to the other, giving arguments at one time in favour of the proposal, and following them with others against. Thus, whatever a man's opinion might be, he got new arguments in its favour and voted as he had intended.

The relation between the University and Colleges was profoundly changed in 1882 by the recommendations of the University Commission, which came into force in that year. They required the Colleges to pay annually a certain proportion of their incomes, to be fixed by the Commissioners, to the University. The percentage was graduated according to the wealth of the College : it could not be less than four nor greater than twenty-one per cent. A great deal of trouble had been taken over fixing this, and though it demanded substantial sacrifices from the Colleges it was not by most people regarded as unreasonable, considering the urgent need of the University for money. It brought, however, into conflict more clearly than had ever been done before, the claims of the University and College. To most members of the University the College made by far the stronger appeal. At that time the majority of the members of the University had made very few contacts with it. When they matriculated they had to pay a fee to the Registrary, and at the various University examinations for their degree they might catch sight of their Examiners, or they might come into contact with the University Proctors, but tax-collectors, Proctors and Examiners are not promising material for exciting ardent affection. On the other hand, they owed to their Colleges a delightful home, their salaries, pleasant society, in many cases their teaching, for the lectures on classics and mathematics were at that time given by College Lecturers in College lecture-rooms. Again, in not a few cases it was a scholarship given by the College that had made it possible for them to come to Cambridge at all. All these things made their affection for the College much warmer than that for the University, and made them opponents to any changes which would seriously affect the prosperity of the College. It was not very long, too, before it became apparent that the great and rapid increase in the number of the new subjects for which the University, if it were to maintain its efficiency, would have to provide new Professorships and new laboratories, would make its income even when supplemented by the amounts contributed by the Colleges still quite inadequate. The most obvious way to increase the income of the University was to demand more from the Colleges, and many were afraid that it would not be very long before an attempt was made to do this. As Sidgwick was probably the most conspicuous of those who put the needs of the University before those of the Colleges, there were a good number who thought that everything he proposed had something lurking inside it, which would make it easier to extract money from the Colleges, and they voted against it. Fortunately, however, soon after the beginning of this century the University began to receive a succession of very handsome bequests and donations, and these, aided by a liberal grant from the Government, have put the finances of the University in such a good position that it has been quite unnecessary to ask for any increase in the contribution from the Colleges. The income of the University from all sources has increased from about £60,000 in 1900 to £212,000 in 1930. It is not a very wild hypothesis to suppose that this has been to a large extent due to the important and very interesting discoveries which have been made in the University, and Cambridge may be quoted as an example of the practical results which come from Research for its own sake.

The University now takes an enormously greater share in the teaching of undergraduates than it did thirty years ago. Now practically all the lectures to Honours students are given by University Lecturers in University buildings. It now, to its great advantage, plays in the student's life a part comparable with that played by his College.

Sidgwick himself was a very generous benefactor of the University, and his gifts were all the more valuable since they were often given on occasions when without them some important work done by the University would have had to be given up, or the opportunity of securing the services of some especially qualified teacher missed. Thus he gave money to continue the teaching of Indian Civil servants which was in danger of being given up, for securing the services of F. M. Maitland as a Reader in Law, for establishing a Professorship of Mental Philosophy, and for the erection of the Physiological Laboratory. Thus they were spread over all departments of University work.

He gave work as well as money to the University. He served for some years on its Council and on the General Board. What was perhaps the most important work of his life (apart from that done as a Professor and writer on Philosophy) was that done in connection with Women's Education. This is not the place to go into detail; suffice it to say that he was the leader and most energetic worker in the movement which, starting in 1870 with lectures to students living near Cambridge, had by 1880 grown to Newnham College with a Hall of Residence, now Sidgwick Hall, in fine grounds and with eighty-five students, and that in 1881 it was granted the privilege that its students should be admitted to the Honours Examinations, and the places they took in the examinations indicated in the class list ; and that after 1880 he and Mrs Sidgwick carried on the work with unabated activity and success. This work will secure him a prominent place in the list of pioneers in Education.

Another subject on which he spent an immense amount of time and work was Psychical Research. He began taking an interest in such subjects when an undergraduate, for then he joined a society called the Ghost Society, founded by Archbishop Benson when he was at Cambridge, and of which it is believed Dr. Westcott was Secretary, for the investigation of ghost stories. Accounts of abnormal experiences such as hallucinations, premonitions, phantasms of the dead and living and those occurring at spiritualistic seances were published from time to time, but no one troubled to test the evidence in support of them. By 1882 two eminent men of science, Sir William Crookes and A. R. Wallace, had expressed their belief in Spiritualism, but by scientific men in general such subjects were regarded as " untouchables " ; anyone touching them would lose caste.

This was most unsatisfactory, for if even a very minute fraction of the things reported were true, the results were of transcendental importance and would revolutionise our ideas about physical as well as spiritual things. The Psychical Society was founded largely through the zeal and energy of Professor W. F. Barrett in 1882 to investigate questions of this kind, and Sidgwick consented to be its first president. He was an ideal president for such a society, absolutely fair and unbiassed and critical. The Society welcomed the communication of accounts of abnormal occurrences and then proceeded to test their evidential value. This involved an immense amount of correspondence. Sidgwick says in his diary that Mrs Sidgwick had gone away on a visit taking with her 1000 scripts on phantasms of the dead. It also led to many bitter disappointments. One night he writes in high spirits in his diary about the joy of discovery, as he thinks he has got conclusive evidence that afternoon of the reality of thought transference at a sitting he has had with a medium in Liverpool. Another sitting the next morning proved that he had been deceived and that the results were worthless.

This work has not been wasted. To put its claims at the very lowest it is surely a great thing to have created an organisation for collecting and testing these abnormal phenomena and thereby to go far to ensure that no genuine ones will escape discovery.

I never heard anything more impressive than the speeches made at a meeting held at Trinity Lodge on November 26, 1900, for the purpose of establishing a memorial to him. There were many speeches, all of them good and all remarkable for the depth of the feelings they expressed. James Bryce and Professor Dicey spoke of the esteem in which he was held at Oxford, and said that there was no one in Cambridge who had had such intimate relations with that university ; his old pupils, F. M. Maitland and James Ward, spoke of what they owed to his teaching and the influence he had had upon their lives ; the Bishop of Bristol (G. F. Browne), who was for a long time the leader of the Conservative party in the University and therefore generally opposed to his policy, spoke of the good work Sidgwick had done in the development of Local Lectures and Examinations; old friends like Dr. Butler, Leslie Stephen and Sir Richard Jebb told us of his triumphs as an undergraduate. What impressed me more than anything else was a sentence in the speech made by Canon Gore, who, balanced precariously on the kerb of the fireplace and apparently oblivious of all the surroundings, said, " We talk in a familiar way about the World, the Flesh and the Devil; one could not know him without thinking that neither the World, the Flesh nor the Devil had any place in him or about him."

James Ward

James Ward, who succeeded Henry Sidgwick as Lecturer in Moral Science at Trinity College, had had a very varied experience before joining the College. He had been articled when very young to an architect in Liverpool. He soon gave that up and went for six years to Spring Hill College, a college for the training of Congregational ministers. He then became the minister at the Congregational Chapel in Cambridge, but resigned after twelve months in consequence of a change in his religious opinions. He was elected to a Trinity Scholarship in 1873, and when in 1875 Trinity offered a Fellowship in Moral Science, he was the successful candidate in an exceptionally strong field. The other candidates were F. W. Maitland, who became Downing Professor of the Laws of England; Arthur Lyttelton, who became Master of Selwyn; and William Cunningham, who became a well-known authority on Political Economy, and advocated, with great ability, views which were then heretical but which re-sembled in many respects those which are now prevalent. He became later Vicar of Great St. Mary's and Archdeacon of Ely. All the candidates had been at the top of the list in one year or another of the Moral Sciences Tripos. Ward was the only one in that class in his year. He was elected to a Lectureship in Moral Science in Trinity College in 1881, and to the Professorship of Mental Philosophy and Logic in 1893. He was a Tutor of Trinity from 1896 to 1897. This was a post for which he was not well fitted. He had not become an undergraduate himself until he was thirty, and knew very little about, and perhaps had but little sympathy with, the views and pursuits of the normal undergraduate.

After his election to a Fellowship he worked in the Physiological Laboratory under Michael Foster, who thought so well of him that he said a good physiologist was lost when he devoted himself to philosophy. Ward even published a paper in the *Journal of Physiology*. He was a severe critic of his own work as well as that of other people. He was never satisfied with what he had done and kept putting off publication in the hope of making it better. His best work was done on commission. Thus Robertson Smith, the editor of the *Encyclopaedia Britannica*, persuaded him to write the article on Psychology, which at once established his reputation as a psychologist of the very first rank. Again, his *Naturalism and Agnosticism* and *The Realm of Ends*, which won for him the same position in other branches of philosophy, were the result of his under-taking to give the Gifford Lectures in the Universities of Aberdeen and St. Andrews respectively. His conversation, though it did not sparkle with epigrams and paradoxes like that of some of his contemporaries, was quite as impressive. One could not talk with him on a serious subject without recognising that he had a mind of quite exceptional acuteness and had thought deeply on everything he spoke about. He was an excellent field naturalist, and a walk with him in the country was most interesting and instructive, for he recognised and had something new to tell one about nearly every bird or flower that came in the way. Though he was rather an austere man he had many warm friends in Cambridge, and the admiration and respect which was felt for him was expressed by the presentation to him on his eightieth birthday of his portrait by McEvoy.

J. McT. E. McTaggart

John McTaggart Ellis McTaggart came up to Trinity from Clifton in 1885. He was alone in the first class of the Moral Science Tripos in 1888, and was elected to a Fellowship at Trinity in 1891. He then paid a long visit to New Zealand, where in 1894 he did the wisest thing he ever did in his life by marrying Miss Margaret Bird. As an undergraduate he was prominent among the " intellectuals " of his year. He was an " Apostle " and a successful speaker at the Union, where he was President in 1890. In 1897 he was made Lecturer in Moral Science at Trinity College, and held this post until he retired after twenty-five years' service in 1923. In addition to his lectures to students of Moral Science, he gave a course of one lecture a week intended for those who were not specialising in philosophy but who wished to get a general idea of what it was about. This course was extraordinarily successful ; it was attended by large audiences and he repeated it year after year. These lectures probably made many take some interest in philosophy who without them would never have thought about it. His lectures and writings were exceptionally clear : he never left you in doubt as to what he meant. His witticisms and paradoxes made them sparkling and exciting ; they were very pleasant to listen to or to read.

Mr H. G. Wells in *The New Machiavelli* introduces him under the name of Codger, and describes very graphically and accurately his appearance, his gait and other characteristics, but for all this the Codger of the novel is not McTaggart. Codger is like one of Mr Wells' Martians, all brain and no heart, one who does not love or hate or grieve, whose interests are confined to unravelling the secrets of Hegel. This is not McTaggart. He, besides being a Hegelian, was an excellent man of business and affairs, and enjoyed this kind of work. His love for his old school, his college and his country was intense. I never knew anyone who had a greater love or veneration for Trinity, or one more anxious to keep up its old customs. He never missed a meeting of the Fellows and always took an active part in the discussion. He served on the Council of the College and on numerous committees ; he dined regularly in Hall, and kept up the old custom of going after Hall to the Combination Room, even when, as sometimes happened, he was the only one who did so.

When the war came McTaggart grasped at every opportunity of helping England to win. To save Cambridge from being bombarded by Zeppelins it was desirable to keep it as dark as possible, so a corps was formed to patrol the streets at night and see that all lights were dimmed. No one was more zealous or persistent in this work than McTaggart, and no one, as I know to my cost, had such a severe standard of dimness. The efforts of himself and his colleagues were successful, for Cambridge was never hit by a bomb though many fell not far away. He felt very strongly that it was everyone's duty to do all they could to help to win the war, and he had no toleration for those who shirked this work, or still worse, thought it their duty by their speeches and writings to incite others to do so. This led to his breaking off friendly relations with some of his oldest friends.

McTaggart was a man who made up his own mind on political matters and was not in the pocket of any political party. In University matters he was a reformer; he supported the abolition of compulsory Greek, and was strongly in favour of the admission of women to the University. In politics he found it difficult to know who to vote for. The old-fashioned Whigs seemed to represent his views, but where are the Whigs to be found? He was a very strong Free Trader. His views on this were much like those of the " Manchester School " of Cobden and Bright—no interference with trade either by tariffs or by regulations about wages : but now there are no Free Traders who do not hold opinions which he detested more than he did tariffs.

He managed to hold opinions which are not usually reconciled. He was not a Christian, and yet for metaphysical reasons a firm believer in Immortality. He was also a strong supporter of an Established Church, because it excited the antagonism of Dissenters and so weakened the influence of religion on the policy of the country, which in his opinion was a very desirable thing. McTaggart was a lover of ceremonial and ritual; he seized every opportunity of wearing the scarlet gown of a D.Litt., and took a great interest in drawing up the list of occasions on which Trinity should exercise its privilege of flying the

J. McT. E. McTAGGART

Royal Standard. He was the least athletic able-bodied man I ever met. He detested games, and when at Clifton, where games were compulsory, had frustrated the attempt to make him play football by lying flat on the ground and refusing to get up.

W. W. Rouse Ball

W. W. Rouse Ball, who for nearly fifty years did excellent work for the College both in administration and in teaching, and who bequeathed to it one of the largest legacies it has ever received, was born in London in 1850. He came up from University College, London, to Trinity as a Minor Scholar in 1870, and was Second Wrangler and 1st Smith's Prizeman in 1874. He was made a Fellow of the College in 1875, and after a short career at the Bar he returned to Trinity as Lecturer in 1878, and lived in Cambridge for the rest of his life. He held the Lectureship until 1905, and was a Tutor of the College from 1893 to 1905. Rouse Ball took a deep and very genuine interest in his pupils, and did much to establish social as well as official relations with them. He often asked them to dinner, and built a billiard-room and a squash-racquets court for their amusement. He also took pains to keep in touch with them after they left Cambridge. He had supported warmly the scheme for " Old Boys' " dinners, which was put into force in 1889. He took an interest in their sports : he was treasurer of the 1st Trinity Boat Club and wrote its history, and gave a challenge cup to be held by the winners of the Inter-Collegiate Athletic Sports. Ball was a good chess player, and had represented Cambridge in the first Oxford and Cambridge Chess Match in 1873. His

Mathematical Recreations, of which ten editions have been published, discusses arithmetical and geometrical puzzles, magic squares, the problem of the 15 schoolgirls, squaring the circle, trisecting an angle, mazes (he had a maze in his garden), cryptograms, ciphers, etc. Anything of the nature of a puzzle or an ingenious toy had a great fascination for him. He had a large collection of those applications of science to frivolous purposes which can be bought for a few pence from the trays of hawkers in London and Paris. To encourage conjuring he founded the " Pentacle Club " in Cambridge, which gives annual performances of an ambitious kind. He was also interested in cat's cradles, which have an ethnological importance, and gave a lecture on them at the Royal Institution. His *Short History of Mathematics,* which had a very large circulation, is a most interesting book. It is very well written, and the mathematics is leavened by accounts of the idiosyncrasies and escapades of the mathematicians. For a long time he had the intention of writing a biography of Sir Isaac Newton, and had collected a considerable amount of material for it. He was prevented by the pressure of other work from completing this, but he embodied part of it in his very able and interesting *Essay on the Genesis and History of Newton's "Principia".* He had formed a very fine collection of portraits of mathematicians. Rouse Ball was interested in the history of Trinity College as well as of mathematics : he knew not only its official history as expressed in minutes of its Governing Body, but, better than anyone else of his time, its more intimate history, the gossip of the Combination Room, epigrams and verses like those of Porson and Mansel, and the ephemeral literature, such as is contained in the fly-sheets which had been issued when some burning question was dividing the College.

He published a very short *History of Trinity College* in Dent's series of histories of Cambridge Colleges. In collaboration with Mr Venn (now President of Queens' College, Cambridge) he edited four volumes containing the names of those admitted to the College between 1546 and 1900.

Mr Ball did a great amount of administrative work for the College: he served on its Council, with only four years' interval, from 1888 to 1925, and was its Secretary from 1891 to 1894. Besides this, he was a member of many other important Committees of the College. He was an excellent man of business, had very sound judgment, never wasted any time at a meeting, stated a case very clearly and fairly, and nothing could ruffle his courtesy. This was a great help in soothing down acrimonious discussions. A Vice-Chancellor, who had to preside at the meetings of the University Finance Committee at a time when it contained three pugnacious Bursars as well as Ball, told me that it was only Ball's efforts which prevented the Committee becoming a bear-garden.

The funds of Trinity College and also of the University owe much to Mr Ball's ability in financial matters. When his father died he inherited a considerable sum of money: he determined to make it the nucleus of a fund for the promotion of education and research, and for this purpose he made it into a trust, the trustee being an important financial house in New York. According to the provision of the Trust, Ball was to have the control of the investments and the power to withdraw money from the Trust, provided the sum withdrawn was devoted to education or research. He devoted, he told me, unremitting attention and much labour to the investment of the funds of the Trust, and was so successful that, at the time of his death, the Trust had increased to many times its original value, even to many times what its value would have been if the interest had just been allowed to accumulate without change of investments. There had been no great coup, but just a steady flow of small increments. By his will he established new Professorships in Mathematics and in Law at both Cambridge and Oxford, and left substantial sums to the Cambridge University Library and also to the Library of Trinity College. The residue, which was the major part of the estate, went to Trinity College. The sum received from this legacy has only been exceeded once, and that by a relatively small amount, in the history of the College. It is unique in that it was due largely to his working for many years with the set purpose of increasing the power of the College to fulfil one of its most important functions—the promotion of research. He had in his lifetime made gifts to the College, one for a fund to enable students to meet medical expenses, and another for a travelling studentship in mathematics, and had founded in the University the Rouse Ball Lectureship in Mathematics.

He was a good friend of mine for nearly fifty years, and I am grateful to him for many kindnesses, much help and wise counsel.

Reginald St. John Parry

Reginald St. John Parry came up to Trinity with a Minor Scholarship in 1876. He was 2nd Classic in 1880, was elected to a Fellowship in 1881 and to an Assistant Lectureship in Classics in the same year. The College was his home until he died, as he never married. His life was spent in the service of the College : he had held every major College appointment, and was Vice-Master

from 1919 until his death in 1935. He took a large part
in the administration of the College, he served many years
on the College Council and on innumerable Committees.
For a long time he was Secretary of the Livings Committee
and took a leading part in the very difficult business of
finding suitable men to fill vacant College livings. He was
a Canon of Southwell Cathedral. Perhaps the most important work he did for the College was through the
"At Homes" he held on Sunday evenings in Term time.
He was very catholic in his invitations to these; he was
especially anxious to include undergraduates who had not
many friends in Trinity before they came up. He was an
excellent host, and managed to get his guests to speak
freely to each other, and many men owe to these At
Homes friendships which they regard as among the best
of the good things they got at Trinity. Meetings of this
kind help to minimise one of the disadvantages of a large
College where men may not meet each other so frequently
as they do in a smaller one. Parry knew a large number
of the undergraduates in residence and kept in touch with
many of them after they went down. One of the attractions to our guests at the Old Members' dinner was the
chance it gave them of meeting him again. When
Henry Jackson's health failed, he took on "Jackson's"
party after Commemoration. Parry was a prominent
supporter of Reform in both College and University
questions ; in fact his opinion on these was generally
much the same as Jackson's. He took a share in University as well as College Administration, served for several
years on its Council, and was for a time Chairman of the
University Press Syndicate. He took a great interest in
"Adult Education", and in 1930 received an address signed
by Mr Stanley Baldwin and a large number of distin-

guished people, expressing their profound sense of the invaluable services which he had rendered for over thirty years to the cause of Adult Education.

Srinivasa Ramanujan

Few, if any, Fellows of Trinity can have had a more romantic career than Ramanujan. His parents were Brahmins, living in the Madras Presidency and in poor circumstances. When he was seven years old he went to the High School at Kumbakonam and remained there until he was sixteen. He showed quite remarkable mathematical ability, and in his last year, after he had access to Carr's *Synopsis of Pure Mathematics*, which contains a list of mathematical formulae without any proofs, he revelled in finding the proofs for himself. He matriculated at the University of Madras and studied at the Government College at Kumbakonam, gaining a scholarship awarded for proficiency in mathematics and English. His University career was a failure. He spent all his time on mathematics, even that in the lecture-room when he was supposed to be listening to lectures on other subjects. Naturally he failed to pass in an examination which included these subjects. Another attempt resulted in another failure. He kept on, however, working at mathematics and entering the results he obtained in his note-books. In 1909 he got married, and it was necessary for him to obtain some post to provide a livelihood. Owing to his failure to get a degree this was difficult, but he finally obtained a subordinate post, with a salary of 30 rupees a month, in the Madras Port Trust Office. Though the salary was so small, it was, as things turned out, the turning-point in his career, for the manager of the Company was a mathematician and took a great interest in Ramanujan. On the advice of his friends, Ramanujan wrote to Mr G. H. Hardy, a mathematical Lecturer and Fellow of Trinity College, giving a list of a number of theorems he had discovered. Some of these were not new, a few were not true, but there were some so important that Mr Hardy was convinced that Ramanujan had mathematical abilities of the highest order, and he started enquires to see if steps could not be taken to enable him to come to Cambridge. Ramanujan was asked if he would go to England, but since he would lose caste by doing so, he said no. Another Fellow of Trinity, Mr (now Sir) Gilbert Walker, who was the head of the Indian Meteorological Office, happened to visit Madras and was shown some of Ramanujan's work. He wrote to the University of Madras suggesting that steps should be taken to enable Ramanujan to devote the whole of his time to mathematics. In consequence of this the University, with the consent of the Government, gave Ramanujan a special scholarship of 75 rupees per month. Soon after this another Fellow of Trinity appeared on the scene, in Mr Neville, who was invited to give a course of lectures on mathematics at Madras University. When Mr Neville got to Madras he set to work to persuade Ramanujan to alter his decision about coming to England. He found that he himself was not unwilling to do so—the difficulty was with his mother, who would not give her consent, and he would not come without it. One morning, however, she said she had had a dream in which she saw her son in a big hall in the midst of a group of Europeans, and that the goddess Namigiri had commanded her not to stand in his way. After this, all was easy. The

University of Madras granted him a scholarship of £250 a year for two years, tenable in England, and Ramanujan, having arranged that his mother should receive 60 rupees per month out of this, sailed for England in March 1914, reached Cambridge in April, was admitted to Trinity College, and given an exhibition of £60 per annum to supplement his scholarship from Madras. When he arrived in Cambridge it was found, as might have been expected from his training, that, while he had a profound knowledge of some part of mathematics, there were many parts of which he was quite ignorant, and some of these were important in connection with the parts of mathematics in which he was interested. He was said also to have a very vague idea of what constituted a mathematical proof. After a few years' teaching by Mr Hardy, these defects were remedied without checking the flow of his original work. Ramanujan published a number of important papers which led to his election to the Royal Society in the spring of 1918, when he was barely thirty. He was the first Indian to be elected and he was elected the first time his name was in the list of candidates, which is somewhat unusual. In the same year he was elected to a Fellowship at Trinity College, which entitled him to an income of about £250 for six years unconditionally.

His health broke down early in 1917, and he went into a nursing home, first in Cambridge, and afterwards in Wells, Matlock and London. Late in 1918 he got distinctly better, and as it was thought that his illness might have been due to the difference between the climate and food of Cambridge and India, he returned to India early in 1919. This was, however, of no avail, for he died on April 20, 1920.

A volume containing his collected papers, edited by

G. H. Hardy, P. V. Seshu Aiyat and B. M. Wilson, was published by the Cambridge University Press in 1927. In addition to this there were in his note-books statements without proofs of a large number of theorems. These have been worked over by several eminent mathematicians, who have succeeded in proving the correctness of a good many of them, and thereby greatly strengthened the verdict of Professor Hardy that in his own field he was unrivalled in his day. His method, however, was not the normal one in which the theorem arises out of the proof; no proof, no theorem. It is possible, however, to imagine other ways of proving theorems. Suppose, for example, a mathematician dreamt that he had discovered a new theorem, if he remembered it when he awoke he might test it by seeing if it gave the right result in a great number of special cases. This is the method of " trial and error "; the great difficulty in this method is to get something to try; there are an infinite number of things which might be tried, and unless we had something to guide us the chance of choosing the right one would be infinitesimal. It need not require a dream to act as a guide. One who, like Ramanujan, had made a long and intense study of a particular branch of mathematics might almost unconsciously have been led to recognise certain features, such as the absence or presence of certain arrangements of the symbols in the theorems known to be true, and would instinctively reject a theorem in which these did not occur. He would, by long experience, have acquired an instinct by which he could distinguish between theorems which were possible and those which were impossible. Then, if he had the imagination to think of a theorem which would satisfy the test, and the industry and power of calculation required to verify it, he might arrive at theorems which he could not prove. There are several instances of mathematical theorems which are believed to be true but have never been proved. Perhaps the most famous is the formula given by Gauss for the number of prime numbers (*i.e.* numbers which, like 2, 3, 5, 7, 11 . . ., are not divisible by any other number), which are less than a given number N ; Gauss' formula was tested for integral values of N up to a thousand millions, and was found to give the right result. This was universally accepted as overwhelming evidence of its truth, though no formal proof had been discovered. Professor Littlewood has shown, however, that it must ultimately fail when N is greater than a certain number. This number, however, is so prodigious that it would be beyond the power of human effort to count by trial the number of primes. Thus Gauss's rule may console itself by thinking that though it may lapse from rectitude it never does so when it can be found out.

A. E. Housman

A. E. Housman became a Fellow of Trinity in 1911, soon after he had been elected to the Professorship of Latin in Cambridge University. After his election to a Fellowship he made the College his home, living in rooms in College and dining regularly in the College Hall. His rooms were for many years on Staircase K, Whewell's Court, up many flights of stairs. To avoid these he moved into ground-floor rooms on Staircase B, Old Court, a few months before he died.

He joined, in 1919, a dining-club of resident Cambridge graduates which met once a fortnight in term time, and of which I was a member, and he was very seldom absent from their dinners. I had thus for nearly twenty-five years many opportunities of meeting him, and after this long experience I think that his silence and aloofness were very much exaggerated in some of his obituary notices. It is true that from time to time he had fits of silence and depression ; but these were rare, surprisingly so, in view of the pessimism of much of his poetry. He usually, in my experience, talked freely and, as might be expected, incisively. He held strong opinions on many subjects and expressed them strongly, and he was not fond of strangers. I always found him excellent company, and was very glad when I could sit next to him.

His appearance and his tastes were very different from those popularly attributed to poets ; he had nothing of the poet about him, except the poetry. He was careful about his dress, which was not marked by any eccentricity, his hair was short. He looked much younger than his years. I was much perturbed one morning when reading *The Times* to see that it was his seventieth birthday, for I had not sent him, as I should have liked to do, my good wishes on such an important event. He took my excuses in very good part, and said that I was not the only one who had misjudged him, for a few days before, when he was walking along a country lane, a farm hand who was driving a cart, coming up from behind, had called out " Hey, you lad, get out of the way ". He liked good food and good wine and was a connoisseur in both ; in fact he was for many years a member of the committee which chooses the wine for the College. Housman took much interest both in wild flowers and in gardening : he was a member of the Garden Committee for many years, and for a part of the time its Secretary. This is a post of con-siderable responsibility, as the secretary is responsible for the supervision of the garden and the gardeners, and for seeing that the recommendations of the committee are carried out. He was a very active and useful member of the committee. He held very decided views, which in general seemed to me quite sound, about the desirability or otherwise of changes which might be suggested. He liked flowers to have bright and definite colours, and was very contemptuous of what are called in florists' catalogues " art shades ", and which he called muddles.

Though he was quite indifferent to distinctions and had refused one which many regard as the greatest which could be conferred upon them, there was one which I think he did appreciate : it was to have had a dish, Barbue Housman, which is a speciality of a famous Paris restaurant, named after him. The dinners which he gave as a member of the Dining Club had, like everything he did, the air of distinction. There was always some dish which few, if any, of his guests had met with before, and over which he had taken a good deal of trouble to instruct the College cook in all the details of its preparation. All the wine was good, and there was pretty sure to be some of special interest or rarity. He recognised, too, the virtues of beer. Has it ever been exalted to such a height of dignity as in

> Malt does more than Milton can
> To justify God's ways to man ?

In the Leslie Stephen lecture, " The Name and Nature of Poetry ", he describes the way his poems had been made : in this, beer plays a part. They were not made by a sad mechanic exercise, but " there would flow into my mind, with sudden and unaccountable emotion, sometimes a line or two of verse, sometimes a whole stanza at once, accompanied, not preceded, by a vague notion of the poem which they were destined to form part of". He found that this flow was helped if he had a pint of beer at luncheon : this acted as a sedative to the brain, and made it more likely to respond to abnormal influences (we may compare it to the trance which precedes hypnotic phenomena). To Housman, poetry was something which excited certain emotions ; it seemed to him more physical than intellectual, and its production differed from that of prose by being passive and involuntary rather than active. It came as it were of itself and not by conscious thought. I remember asking him once if he thought it was possible that Tennyson's " Crossing the Bar " could, with its exquisite phrasing, have been composed, as has often been stated, in its final form in the forty minutes or so while the poet was crossing from the mainland to the Isle of Wight. He said he thought it was quite possible, for if the poet had one of these fits of inspiration the right words would come of themselves.

Though he could write dignified and vigorous prose such as few could equal, he disliked doing so ; it did not, like his poems, come spontaneously, and he had to spend much work and time before he got it into a form which satisfied him. It required a great deal to do this, as he was very fastidious. He was never satisfied with a thing that was good only in parts. In his lecture, Housman quoted passages from Shakespeare as examples of supreme poetry, but I have heard him say that it gave him no pleasure to read a play of Shakespeare's from beginning to end, for though some parts were magnificent, there were others so slovenly that the effect of the whole was disagreeable.

Housman was not only a poet, he was also Professor of Latin, and at the conclusion of his lecture on poetry he said that was his proper job. I am quite incompetent to form any opinion of his merits as a classical scholar, but I know that many high authorities regard him as the greatest England has produced since Bentley ; even laymen can verify that he was as vigorous as that great man in his criticism of those who differed from him. Housman did once wander into a subject which is at any rate closely akin to mathematics. He studied astrology when he was preparing his edition of Manilius, and learned how to cast horoscopes. Astrology is closely connected with the motion of the planets, and thus involves ideas which are sufficiently mathematical to scare off the great majority of classical scholars. His lectures were, I am told, confined to the text of the author he was considering and he did not discuss its literary merits. The one exception, I believe, was when he lectured on Horace, when he gave a translation of *Odes*, iv. 7, into English verse, and was so much moved by it that his eyes filled with tears.

He continued to give his lectures even after his health broke down, sometimes coming from the nursing-home to the lecture-room, and going back there as soon as the lecture was over.

I saw him on the day he gave his last lecture. He was terribly ill and must have had invincible determination to lecture in such a state. He was taken to the nursing-home the next day, and died there on April 30, 1936. The funeral was in the College chapel, and the hymn was one he had written himself and sent a year before to the Dean of Chapel, asking that it might be sung at his funeral.

I have not space enough to give an account of more than a minute fraction of the number of Trinity men I have known during my sixty years' residence in Cambridge. The few I have given have been in the main those who had very long connections with the College and were known to many generations of Trinity men, or were connected with events of exceptional interest. I feel, however, that I should be underrating my debt to the College if I left without mention, however brief, some others whose friendship I owed to my connection with Trinity.

G. F. Cobb, who had been alone in the First Class of the Moral Sciences Tripos in 1861, became Junior Bursar in 1869. He was a good musician and his settings to music of several of Kipling's *Barrack-Room Ballads* became very popular. When he was a young man he was what would now be called an Anglo-Catholic, and wrote a book called the *Kiss of Peace*, advocating closer union with the Roman Church. This aroused the ire of another Fellow of the College, Sedley Taylor, who was in the same year as Cobb and was also a good musician, but he was also a mathematician and could not forgive the treatment of Galileo, so he wrote a reply called the *Kick of War*. He wrote also an excellent book, *Sound and Music*, and invented an instrument called a phoneidoscope, in which soap films showed beautifully coloured patterns when musical notes fell upon them. He was an excellent teller of good stories and no one enjoyed them more than himself; after telling the story he rubbed the palms of his hands vigorously together and beamed on all around him. He was interested in social questions and found the money for providing periodic inspection of the eyes of children in Cambridge schools. He left a large bequest to the College, and an important road in Cambridge is called the Sedley Taylor Road.

His namesake, H. M. Taylor, who had been Mathematical Lecturer and Tutor, was remarkable for the resolution and success with which he fought against the disabilities imposed by blindness which came upon him when he was middle-aged. He continued to take an active interest in municipal affairs. He became Mayor and fulfilled efficiently all the duties of that office. He even, after he was blind, brought out for the Pitt Press an edition of Euclid. It is remarkable that R. D. Hicks, a very learned Greek scholar and a Fellow and Lecturer of the College, also became blind, from the same cause, " detachment of the retina ", as Taylor, and only a few months afterwards.

A. W. Verrall, a lecturer in Classics and at one time Tutor, was a very brilliant and stimulating lecturer, and delighted large audiences by his ingenuity in suggesting emendations to obscure passages in classical authors. His colleagues were not always convinced of the soundness of some of these, but it is certain that he was exceptionally successful in arousing the interest of his hearers, which is one of the most important things a lecturer can do. He held the Clark lectureship in English Literature for one year, and the lectures he gave were exceptionally successful.

F. J. H. Jenkinson, also at one time a lecturer in Classics and later the University Librarian, and in addition an excellent field naturalist, was one of the best beloved Trinity men of his generation. There is an admirably vivid and intimate biography of him by his brother-in-law, Dr. H. F. Stewart.

J. P. Postgate, a distinguished Latin Scholar, was also a Classical Lecturer in Trinity from 1884 to 1909. He was one of the pioneers in introducing what was then the new pronunciation of Latin. I was once a victim of this. When I was admitted to the Professorship, I had by the regulations to make a declaration in Latin before the Vice-Chancellor. Postgate, who was a great friend of mine, asked me to use the new pronunciation. I said I had no prejudices one way or the other, and if he would teach it me I would use the new. I began doing so, but I had not gone very far before the Vice-Chancellor, who was a mathematician, stopped me, and said that by the regulations of the University I must make the declaration in *Latin*. So I had to begin again, and pronounce it as I should have done if I had been left to myself.

James Stuart, who played a very prominent part in University affairs between 1870 and 1890 and was the originator of important developments of University activities, became a Fellow in 1867 and was a Lecturer in Mathematics from 1868 to 1875, when he became Professor of Mechanism. He was a pioneer in the establishment and organisation of lectures and classes outside Cambridge, and it was his enthusiasm and persistence which induced the University to undertake this work. It was very successful; in 1913 lectures were given in forty-nine places outside Cambridge. When he was elected to the Professorship of Mechanism, he began by taking steps to obtain a workshop and drawing office for the instruction of his students, and these, on a very modest scale, were established by the end of 1881. Mr J. A. Fleming, now Sir Ambrose Fleming, F.R.S., was appointed Demonstrator. The number of engineering students rapidly increased, and when he resigned the Professorship in 1890 and was succeeded by Professor Ewing, they were numerous enough to justify the establishment of an Engineering Tripos. Thus it was Stuart who started the Engineering School in Cambridge.

George Howard Darwin, the son of Charles Darwin, was elected Fellow of Trinity in 1868; his Fellowship expired in 1878, but he was re-elected in 1883 when he became Plumian Professor. Except for a short time after his first election to a Fellowship, he worked uninterruptedly in Cambridge on questions such as the genesis of the moon, the theory of tides, periodic orbits, all alike involving mathematical investigations which were not only difficult but very lengthy, and required great patience as well as great skill. This may be regarded as the apotheosis of arithmetic, as so much of it consisted of arithmetical calculations. Lord Kelvin quickly recognised the importance of his work, encouraged him to go on with it, and a warm friendship sprang up between them. Darwin himself was very generous in the encouragement he gave to young men, as I can testify by personal experience. His work was recognised by many scientific societies both at home and abroad. He was President of the Astronomical Society and received their Gold Medal. He was President of the British Association when it met in South Africa. He received the Royal and the Copley Medals from the Royal Society and it is an open secret that he would have been invited to accept the Presidency in succession to Sir Archibald Geikie, who would retire on December 1, 1913. His unfailing courtesy and his interest in many branches of science would have made him an ideal President. But it was not to be. His health had been bad from the spring of 1912, and in the autumn it was found that he was suffering from a malignant disease from which he died at the age of sixty-seven on December 7, 1912.

John Willis Clark, who was a Fellow from 1858 to 1868, was a very conspicuous figure in University life for more than fifty years. He was a native of Cambridge, where his father was Professor of Anatomy. He had so many and such varied activities, and had held so many offices, including that of Registrary of the University, that it is impossible in a short notice to give any adequate account of them. Fortunately this is not necessary, for there is a most entertaining, interesting, sympathetic and intimate biography written by his close friend, the late Sir Arthur Shipley, F.R.S., Master of Christ's College.

John Newport Langley, F.R.S., graduated from St. John's College but was elected to a Trinity Fellowship in 1877, and was a Lecturer from 1884 to 1903, when he succeeded Michael Foster as Professor of Physiology. By his own fundamental discoveries on the nervous system and also by the good work done by his pupil, he increased the already great reputation of the Cambridge Physiological School, and it was under his active supervision that the new Laboratory was erected. He was also for long a most efficient Editor of the *Journal of Physiology*. He had many interests and accomplishments; among other things he was one of the best skaters in England in the old style, which was more "swanlike" than that now in vogue, though not so acrobatic. He was a good man of affairs, a good man on a committee and a good companion.

Walter Morley Fletcher, who became a Fellow in 1897, a Lecturer in 1899 and was Tutor from 1905 to 1915, was a pupil of Langley's, and there were points of resemblance between them, especially in the width of their activities. Thus Fletcher, while an undergraduate, got First Classes in both parts of the Natural Sciences Tripos, and also his "Blue" for athletics. Besides his scientific activities, Fletcher took an interest in art. He was very active and helpful in regard to schemes involving altera-tions in the structure or appearance of the College. He helped Langley in the design of the new Physiological Laboratory, and the undergraduates in planning extensive alterations to the Pitt Club. Throughout his life he took a great interest in athletics and, indeed, in sport of all kinds. His most important work was done after he ceased to be Tutor in 1915. In that year he was appointed Secretary to the Medical Research Council, a body with a large grant from Government which had just been formed for the promotion of medical research in this country. The members of this Council had to develop the methods by which this could best be done, and in such a case as this a great deal depends on the energy and powers of organisation of the Secretary. Fletcher was eminently successful, and by the time of his death in 1934, the Medical Research Council was stated by an authority as high as that of the President of the Royal Society to have brought medical research in this country to a level comparable to that in any other. He had a wide circle of warm friends of all sorts and conditions ; he was a Fellow of the Royal Society and was made a knight in 1918.

Anthony Ashley Bevan was elected a Fellow in 1890. He remained a Fellow until his death, but refused to receive any dividends. He was made a Lecturer in Hebrew and Oriental Languages in 1887, and Lord Almoner's Professor of Arabic in 1893. Hebrew and Arabic were not the only languages in which he was proficient. He had been educated in Lausanne and spoke French and German as fluently as he did English. Whenever there was a guest in Hall who could not speak any English we tried to arrange that he should sit next Bevan. He had charming manners, tinged with a courtliness which was almost French. Though there were but few students taking Hebrew or Arabic, he managed to make contact with many undergraduates, and often invited them to his rooms after Hall and gave them very hot chocolate to drink. He was a very generous man and a liberal benefactor of the College.

SIR J. J. THOMSON

from the bust by F. Derwent Wood in the Library of Trinity College,
Cambridge.

CHAPTER XI

Discharge of Electricity through Gases; The Discovery of the Electron; Positive Rays

IT was a most fortunate coincidence that the advent of research students at the Cavendish Laboratory came at the same time as the announcement by Röntgen of his discovery of the X-rays. I had a copy of his apparatus made and set up at the Laboratory, and the first thing I did with it was to see what effect the passage of these rays through a gas would produce on its electrical properties. To my great delight I found that this made it a conductor of electricity, even though the electric force applied to the gas was exceedingly small, whereas the gas when it was not exposed to the rays did not conduct electricity unless the electric force were increased many thousandfold. It required an electric force of about 30,000 volts per centimetre to make electricity pass through air at atmospheric pressure when not exposed to the rays, while a minute fraction of this was sufficient when the rays were going through it. This was a matter of vital importance for the investigations on the passage of electricity through gases, a problem at which I had been working for many years. Until the rays were discovered the only ways of making electricity pass through a gas were either to apply very great electric forces to it, or else to use very hot gases such as flames. In either case it was exceedingly difficult to get anything like accurate measurements. The results were apt to be very capricious, apparently depending upon causes which it was very difficult to locate. We know now that these two phenomena are the most complex and difficult in the whole range included in the subject of conduction of electricity through gases. To have come upon a method of producing conductivity in the gas so controllable and so convenient as that of the X-rays was like coming into smooth water after long buffeting by heavy seas. The X-rays seem to turn the gas into a gaseous electrolyte. Indeed, many of the difficulties which had been met with in the early stages of the conduction through liquid electrolytes had been met with in those of the conduction through gases. One of these difficulties in the case of liquid electrolytes was to see how the positively and negatively charged atoms in a molecule of the electrolyte could be separated from each other by the very small electric forces which are adequate to produce conduction. The attraction between these oppositely charged atoms is of the order of 1,000,000,000 volts per cm., and yet a force of a fraction of a volt per centimetre is sufficient to make an atom with a positive charge appear at one electrode and one with a negative charge at the other. To explain this Grotthus, in 1805, introduced the idea that the molecules with their + and − atoms formed themselves into chains, like iron filings when acted upon by a magnet and that the + atom in one molecule went close to the − atom of the adjacent molecule in the way indicated in the figure. AB, CD, EF represent molecules.

$$\left(\begin{smallmatrix} A & B \\ - & + \end{smallmatrix}\right) \left(\begin{smallmatrix} C & D \\ - & + \end{smallmatrix}\right) \left(\begin{smallmatrix} E & F \\ - & + \end{smallmatrix}\right)$$

The attraction of the positive charge of B on the atom A is balanced by the repulsion due to the negative charge on C, which is close to B. Each atom, except those at the end, has a close neighbour of opposite sign, so that the two balance each other. Thus the forces exerted by all the atoms except those at the end of the chain are balanced, and the forces on those at the end correspond to the forces between two particles separated by the length of the chain, instead of by the distance between two atoms in one molecule. A very short chain would be all that was needed to make the attraction between the oppositely charged atoms at the ends so small that an infinitesimal electric force would be sufficient to separate them. The view—now generally accepted—introduced by Arrhenius, that the molecules in a solution are dissociated by the action of the solvent, and not by that of any external electric force, has made the introduction of the conception of Grotthus' chains unnecessary. A difficulty of the same type was met with in the study of the passage of a current of electricity through a gas under the action of an electric force between electrodes immersed in it. There was very strong evidence to show that in this case there was decomposition of some of the molecules when the electric force exceeded about 30,000 volts per cm. for air at atmospheric pressure ; this force, though considerable, is infinitesimal in comparison with the 1,000,000,000 volts per cm. required to separate one charged atom from another in a molecule. How, then, are the positively and negatively charged particles separated ? One way which suggests itself is the formation of Grotthus' chains. In the light of what we know now it seems that in a gas, in which there are no free charged particles, and which is protected against anything passing through it from outside, there would be no separation of charged particles unless the electric force had the larger value. The separation of the positively and negatively electrified parts of a molecule is not regarded now as due to the electric force pulling the two parts asunder in a kind of electrical tug-of-war. The rôle of the electric force is rather to give to an extraneous charged particle so much energy that when it strikes against a molecule one part of the molecule is knocked away from the other by the collision. This produces two new free charged particles ; these under the electric field will acquire enough energy to ionise other molecules and so the number of charged particles produced will increase in geometrical progression. The energy that was in the electric field is used up in giving energy to the charged particles. It is a very common experience that when the gas has been very carefully dried, and the walls of the vessel in which it is contained and the electrodes carefully freed from absorbed gases, it is very difficult to get the first spark to pass through it, but when a spark has passed, succeeding sparks pass quite easily. This is because each spark produces many charged particles, and some of these linger in the gas long after the spark has passed. I have known more than one case when to get the first spark to pass it was necessary to increase the electric force to such an extent that, when it did pass, the current was so great that it fused part of the apparatus. When the gas is traversed by Röntgen rays the smallest electric force is sufficient to send a current through it. It does for the gas much what is done by the solvent for the salts dissolved in liquid electrolytes.

I started at once, in the late autumn of 1895, on working at the electric properties of gases exposed to Röntgen rays, and soon found some interesting and suggestive results. The conductivity produced by the rays does not reach its full value the moment the rays start, nor does it disappear the moment they stop. There is an interval when the gas conducts though the rays have ceased to go through it. We studied the properties of the gas in this state, and found that the conductivity was destroyed when the gas passed through a filter of glass wool.

A still more interesting result was that the conductivity could be filtered out without using any mechanical filter by exposing the conducting gas to electric forces. The first experiment shows that the conductivity is due to particles present in the gas, and the second shows that these particles are charged with electricity. The conductivity due to the Röntgen rays is caused by these rays producing in the gas a number of charged particles.[1]

Another interesting point we found when studying the relations between the potential difference between the electrodes and the current through the gas, was that though when the potential difference was small, the current was proportional to the potential difference, as it is in conduction through metals and electrolytes ; yet, as the potential difference increased, the current did not increase in proportion. The increase got smaller and smaller as the potential difference increased, until the current became constant, and did not increase again until the potential difference was so large that it would have produced a current without the aid of the rays. This is just what we should expect if the current is carried by charged particles produced by the rays ; when the current passes these come up against the electrodes, lose their charges, and can no longer carry the current. The current passing through the gas is proportional to the number of charged particles which strike against the electrodes in one second. As

[1] The photographs taken by C. T. R. Wilson many years after the discovery of the conductivity produced by Röntgen rays show that the primary effect of these rays is to eject electrons moving at high speeds from the molecules hit by the rays. These swift electrons eject other electrons from the molecules they hit. Thus the negatively electrified particles start as electrons which finally get attached to molecules.

the charged particles are produced by the rays, it is evident that the number which disappear in one second cannot be greater than the number produced by the rays in that time. Thus the current, when the charged particles are produced by the rays, must have a limit proportional to the intensity of the rays. The measurement of this limiting current or saturation current, as it is usually called, is a very convenient method of measuring the intensity of the Röntgen rays.

The disappearance of the conductivity after the rays are stopped is due to the combination of the positive and negative ions to form a neutral particle, which plays no part in carrying the current. If there are respectively n_1 and n_2 positively and negatively charged particles in unit volume of the gas, the number of collisions occurring in unit volume per unit time is proportional to $n_1 n_2$. Let it be equal to $a\,n_1 n_2$; a is called the coefficient of recombination, and may depend on the pressure, temperature and character of the gas. The current through the gas is due to the motion of charged particles, and will depend upon the velocity they acquire under the action of the electric force. It follows from the kinetic theory of gases that the velocity of a particle moving through a gas is, when the pressure is not very low, proportional to the force acting upon it. If X is the electric force and e the charge on the particle, the force on the particle is Xe; let $k_1 X$, $k_2 X$ be the velocities of the positively and negatively charged particles respectively; k_1 and k_2 are called the mobilities of the positive and negative ions respectively.

Supposing, as we are led to suppose by the facts which have just been stated, that the current is carried by electrified particles which are continually being produced by the Röntgen rays, and that these are then driven against the electrodes by the electric force acting in the gas, we can find a differential equation which expresses the relation between the current through the gas and the potential difference between the electrodes. This, except when the potential difference is very small, is not nearly so simple as that expressed by Ohm's Law. The relation involves the coefficient of recombination a, the mobilities k_1 and k_2, and q, the number of ions per second produced by the rays. The latter is easily determined, since it is measured by the maximum current that can be made to pass without sparking through the gas; a, k_1, k_2, are the fundamental quantities on which the relation depends, and for the next three or four years much of the work of the Laboratory was concentrated on studying these. Rutherford, who had completed his work on the detection of electric waves by a magnetic method, determined the value of a for different gases and made many determinations of the mobilities of the ions by methods of his own devising. Zeleny also made many important experiments on the mobility of ions in different gases by finding the electric force which would just force an ion against a current of gas of known velocity, and showed that the mobility of the negative ion was often greater than that of the positive. He found that this difference was greater in carefully dried and purified gases than in damp or impure gases. Experiments made many years later by Franck show that in certain cases a very small amount of impurity produces an enormous decrease in the mobility of the negative ion. Thus, in pure helium the mobility of the negative was 100 times that of the positive, and the addition of a trace of oxygen reduced this ratio to 1·2. In the pure helium the electrons ejected by Röntgen rays do not attach themselves to the atoms of helium, and so remain electrons with a high mobility. If there is any oxygen present the electrons are captured by the oxygen molecules. The carriers of the negative electricity are then ponderous molecules instead of light electrons. Townsend also worked on the diffusion of ions. Indeed, after a few years' work the properties of the gaseous ions were known with greater precision than the properties of the ions in liquid electrolytes, which had been studied for a very much longer time. An immense amount of work leading to many important discoveries has been done since these experiments were made, but I think the workers in the Cavendish Laboratory may claim to have been the pioneers.

During the period 1896–1900, 104 papers were published by workers in the Cavendish Laboratory. I have not space even to give their titles: one, however, I must refer to. It was the one in which C. T. R. Wilson showed that an electrified body lost its charge in dust-free air, when he had arranged the experiment so that a defect in the insulation of the body would diminish the leak. This occurred when there were no Röntgen rays passing through the gas, and when it was shielded by thick metal from radiation from outside. It was the study of this " residual leak ", so called because every known source had been eliminated, that led to the discovery of the " cosmic rays ". It is one of the romances of science that the study of these very minute, and, what might seem trivial, effects should have led to results which threw much light on a subject of such great importance as the structure of the atom.

Discovery of the Electron

The research which led to the discovery of the electron began with an attempt to explain the discrepancy between the behaviour of cathode rays under magnetic and electric forces. Magnetic forces deflect the rays in just the same ways as they would a negatively electrified particle moving in the direction of the rays.

A Faraday cylinder placed out of the normal path of a thin beam of cathode rays does not receive any charge of electricity, but it receives a copious negative one when the beam is deflected by a magnet into the cylinder. This would seem to be conclusive evidence that the rays carried a charge of negative electricity, had not Hertz found that when they were exposed to an electric force they were not deflected at all. From this he came to the conclusion that they were not charged particles. He took the view which was held by the majority of German physicists that they were flexible electric currents flowing through the ether, the negative electricity flowing out of the cathode and the positive into it, and that they were acted upon by magnetic forces in accordance with the laws discovered by Ampère for the forces exerted on electric currents.

Such currents would give a charge of negative electricity to bodies against which they struck. They would be deflected by a magnet in accordance with Ampère's laws. They would not be deflected by electric forces. These are just the properties which the cathode rays were for a long time thought to possess.

It ought to be pointed out that Goldstein, who for more than thirty years worked indefatigably on cathode rays and made many important discoveries about them, obtained in 1880[1] an effect which we see now must have been due to an effect produced by electric force on the rays. He used a tube in which there were two cathodes side by side, and he observed that when only one cathode

[1] *Eine neue Form elektrischer Abstossung.* Berlin 1880. J. Springer.

was in action the cathode rays from it were not in the same direction as when both cathodes were in action simultaneously. The deflection was in the direction which indicated a repulsion between the two beams of rays.

My first attempt to deflect a beam of cathode rays was to pass it between two parallel metal plates fastened inside the discharge-tube, and to produce an electric field between the plates. This failed to produce any lasting deflection. I could, however, detect a slight flicker in the beam when the electric force was first applied. This gave the clue to what I think is the explanation of the absence of the electric deflection of the rays. If there is any gas between the plates it will be ionised by the cathode rays when they pass through it, and thus produce a supply of both positively and negatively electrified particles. The positively charged plate will attract to itself negatively electrified particles which will neutralise, in the space between the plates, the effect of its own positive electrification. Similarly, the effect of the negatively electrified plate will be neutralised by the positively electrified particles it attracts. Thus charging up the plates will not produce an electric force between them ; the momentary flicker was due to the neutralisation of the plates not being instantaneous. The absence of deflection on this view is due to the presence of gas—to the pressure being too high —thus the thing to do was to get a much higher vacuum. This was more easily said than done. The technique of producing high vacua in those days was in an elementary stage. The necessity of getting rid of gas condensed on the walls of the discharge tube, and on the metal of the electrodes by prolonged baking, was not realised. As this gas was liberated when the discharge passed through the tube, the vacuum deteriorated rapidly during the discharge, and the pumps then available were not fast enough to keep pace with this liberation. However, after running the discharge through the tube day after day without introducing fresh gas, the gas on the walls and electrodes got driven off and it was possible to get a much better vacuum. The deflection of the cathode rays by electric forces became quite marked, and its direction indicated that the particles forming the cathode rays were negatively electrified.

This result removed the discrepancy between the effects of magnetic and electric forces on the cathode particles : it did much more than this, it provided a method of measuring v, the velocity of these particles, and also m/e, where m is the mass of a particle and e its electric charge.

The mechanical force exerted by an electric force X on the particle is equal to Xe and is in the same direction as X ; the mechanical force on the moving particle exerted by a magnetic force H is not in the direction of H, but at right angles to it, and it is also at right angles to the velocity of the particle ; if the magnetic force is at right angles to the direction of motion of the particle the mechanical force is Hev. If the applied electric force X is at right angles both to the velocity and to the magnetic force, the mechanical forces due to the electric and magnetic forces are in the same straight line, and by altering X and H can be made to balance each other, leaving the cathode particle undeflected. When this is the case $Xe=Hev$, so that $v=X/H$. As we can measure easily both X and H, this gives a simple method of finding v. When we have got v, we can get m/e, for a magnetic force applied in the proper direction bends a thin pencil of cathode rays into a circle whose radius is mv/He ; this can be measured, and hence by the magnetic deflection alone we can determine mv/e. This had been done by Schuster ten years before I made my experiments. The magnetic deflection alone does not determine either v or m/e. Schuster assumed that, before reaching the place where the deflections were observed, the cathode particles had made so many collisions with the particles of the gas through which they were passing, that they had been brought into statistical equilibrium with the surrounding gas, and had the same energy as molecules of the gas at the temperature of the discharge tube. On this assumption he came to the conclusion that the masses of the cathode particles were of the same order as the masses of the molecules of the gas through which they were passing. His argument would have been quite sound if the cathode particles were atoms or molecules of the gas, and when he found that on this assumption the magnetic deflection agreed with his observations, the evidence in favour of cathode particles being charged atoms or molecules seemed very strong. I did not see how to reconcile the sharp outline of a thin beam of cathode rays with the idea that its particles had made enough collisions to bring them into statistical equilibrium with the surrounding gas, and it seemed to me necessary to measure directly either the energy or the velocity of the particles. I began by measuring their energy by allowing a pencil of the rays to fall through an opening in a Faraday cylinder which was connected with an electrometer. Inside the cylinder there was a thermopile which the particles heated up when they struck against it. The temperature to which the thermopile was raised in a given time was measured, and, since the heat capacity of the cylinder and its contents was known, the energy communicated in one second could be determined. If n is the number of cathode particles striking the thermopile per second, m the mass, e the electric charge and v the velocity of a particle, the energy, E, given to the thermopile per second is given by the equation $E = \frac{1}{2}nmv^2$; Q, the charge given to the electrometer, is given by $Q = ne$: from the magnetic deflection we get T, where $T = e/mv$: from these equations we find

$$\frac{e}{m} = \frac{2ET^2}{Q},$$

giving the value of e/m in terms of quantities which can be measured. Another method I used was to balance the deflection produced by a magnetic force H against one produced by an electric force X. Then since $v = X/H$, hence

$$\frac{e}{m} = \frac{TX}{H},$$

another expression for e/m.

Using one or other of these methods, I determined the value of e/m for different gas fillings of the tube, air, hydrogen, carbonic acid, but found that the value was the same for each of these gases ; the mean value for twenty-six experiments was $2 \cdot 3 \times 10^7$. The values of v varied in the different experiments from $2 \cdot 3 \times 10^9$ to $1 \cdot 2 \times 10^{10}$ cm./sec.

These experiments were of an exploratory nature ; the apparatus was of a simple character and not designed to get the most accurate numerical results. It was sufficient, however, to prove that e/m for the cathode ray particles was of the order 10^7, whereas the smallest value hitherto found was 10^4 for the atom of hydrogen in electrolysis. So that if e were the same as the charge of electricity carried by an atom of hydrogen—as was subsequently proved to be the case—m, the mass of the cathode ray particle, could not be greater than one-thousandth part of the mass of an atom of hydrogen, the smallest mass hitherto recognised. It was also proved that the mass of these particles did not depend upon the kind of gas in the discharge-tube. These results were so surprising that it seemed more important to make a general survey of the subject than to endeavour to improve the determination of the exact value of the ratio of the mass of the particle to the mass of the hydrogen atom. It was found then, when the aluminium electrodes which had been used in the first experiment were replaced by platinum or other metals, that no effect was produced on the value of e/m, nor did altering the kind of glass used for making the discharge-tube produce any change.

I next tested electrified particles which had been produced by methods in which no electric force had been applied to their source. It is known that metals when exposed to ultra-violet light give off negative electricity, and that metallic and carbon filaments do so when incandescent. I measured, by methods based on similar principles to those used for cathode rays, the values of e/m for the carriers of negative electricity in these cases, and found that it was the same as for cathode rays.

After long consideration of the experiments it seemed to me that there was no escape from the following conclusions :

(1) That atoms are not indivisible, for negatively electrified particles can be torn from them by the action of electrical forces, impact of rapidly moving atoms, ultra-violet light or heat.

(2) That these particles are all of the same mass, and carry the same charge of negative electricity from what-

338

ever kind of atom they may be derived, and are a constituent of all atoms.

(3) That the mass of these particles is less than one-thousandth part of the mass of an atom of hydrogen.

I at first called these particles corpuscles, but they are now called by the more appropriate name " electrons ". I made the first announcement of the existence of these corpuscles in a Friday Evening Discourse at the Royal Institution on April 29, 1897, of which an abstract was published in the *Electrician*, May 21, 1897. It was published at length in the *Philosophical Magazine* for October 1897. About the same time other investigations of m/e were published by Wiechert and Kaufmann, who obtained values which agreed fairly well with mine. They did not, however, make direct measurements of anything but the magnetic deflection of the cathode ray. This only gives mv/e ; they did not make any direct measurement of v or of the energy $mv^2/2$ at the place where they measured the magnetic deflection, but, like Schuster, estimated the energy by making assumptions about the connection between the energy of the particle at the place where the magnetic deflection was measured, and the energy which it would acquire by falling through the potential difference between the electrodes in the discharge-tube. But whereas Schuster assumed that, by collisions with the molecules of the gas, the cathode particle had lost practically all the energy it had acquired from the electric field in the tube, Wiechert and Kaufmann assumed that it had lost none of this energy.

It is not possible to estimate from the potential difference between the cathode and anode of the discharge tube the energy possessed by a charged particle at any point in its course without knowing more about the mechanism of the discharge than we do even at the present time. It is not even possible to determine the limits between which this energy must lie. Thus, for example, a charged particle starting from the cathode would ionise the gas through which it passed and produce other charged particles. These again would produce other charged particles when energised by the electric field, and these again would produce other charged particles. These secondary particles would not acquire as much energy as they would have done if they had started from the cathode itself. Again, if charged particles are liberated from the cathode by the impact of radiation such as ultra-violet radiation or Röntgen radiation of very short wave-length, which is known to be present in the discharge-tubes, the particles would start from the cathode with a considerable amount of energy, and their energy at any place would be greater than that due to the electric field. If negative particles were emitted from the cathode by the impact of positive ions, they might start with energy derived from these particles, and not from the electric field. Measurements of the magnetic deflection only give $v \times (m/e)$. To obtain the value of (m/e) it is necessary to measure v, or some combination of (m/e) and v other than their product.

The interpretation I put upon my results was quite different from that put by the German physicists on theirs. Kaufmann (*Wiedemanns Annalen*, 61, p. 552) interpreted the fact that the value of e/m does not depend on the kind of gas in the tube or on the metal used for the cathode as proving that the negatively electrified particles did not come out of either the gas or the metal. I took the view that all atoms contained these small particles, and that they could be knocked out of the atom by electrical means, by high temperatures or by Röntgen rays.

At first there were very few who believed in the existence of these bodies smaller than atoms. I was even told long afterwards by a distinguished physicist who had been present at my lecture at the Royal Institution that he thought I had been " pulling their legs". I was not surprised at this, as I had myself come to this explanation of my experiments with great reluctance, and it was only after I was convinced that the experiment left no escape from it that I published my belief in the existence of bodies smaller than atoms. There were, however, a few, I think Professor G. F. Fitzgerald was one, who thought I had made out a good case. I went on with my experiments. I determined m/e for the carriers of negative electricity emitted by metals exposed to ultra-violet light ; it was the same as that for the cathode rays. I found also that this was true for the carriers of negative electricity given out by hot metals. I also determined the value of e for the electric charge carried by these negatively electrified particles, and found that it was the same as that carried by the hydrogen atom in the electrolysis of liquids. This left no doubt that the large value of e/m was due to the smallness of the mass and not to the magnitude of the charge. I brought these results before the meeting of the British Association at Dover in 1899, when my great friend Professor J. H. Poynting was secretary of Section A, and I think I made a good many converts to these views.

The Value of e

The work of C. T. R. Wilson in the Cavendish Laboratory on the formation of fogs supplies a method for finding this. Aiken in 1880 discovered that whereas a cloud is produced in ordinary air, saturated with water vapour, when it is cooled by a very slight expansion of its volume, no fog is produced even by very considerable expansions when the dust is filtered out of the air. The drops of water require nuclei on which they can be deposited : in ordinary air these are supplied by the particles of dust which it contains. Wilson found that when the dust-free gas which would not give a fog was ionised, and therefore contained both negatively and positively electrified particles, a fog was produced by an expansion too small to produce one in dust-free air, and that a fog would be produced on negative particles by an expansion less than that required for positive ones. Thus with certain expansions all the drops will be negatively electrified.

Thus if the gas is ionised, say by radiation from radium, it will contain negatively electrified particles ; drops of water will be deposited on these by a supersaturation which is not sufficient to produce any drops when the particles are not electrified, or electrified only with positive electricity. This discovery has been developed into a method which has had a profound influence on the recent study of atomic physics.

If we supersaturate the gas by cooling it by a sudden expansion of known amount, we can calculate the lowering of temperature, and hence the difference in the amount of water required to saturate the air before and after expansion. This will be the amount of water deposited as drops on the negatively charged particle. These drops will fall down slowly under gravity. They will in consequence of the viscosity of the air fall with uniform velocity. This velocity depends on the size of the drops, which can be determined by Townsend's method of measuring the rate at which they fall. We know the total quantity of water deposited on the drops, and since we know the quantity of water in each drop we can find at once the number of drops.

If we drive by an electric force all the negatively electrified particles to a metal plate connected with an electrometer, we can determine the sum of the charges on the negatively electrified particles ; and since the number of these particles is equal to the number of drops, we can determine the charge *e* carried by each particle. A simpler method was invented later by H. A. Wilson. We measure, in the same way as before, the size and therefore the weight of the drop by observing the rate at which it falls. Then we apply an electric force X, which, if *e* is the charge on the drop, exerts an upward electric force X*e*, so that if *w* is the weight of the drop, the vertical downward force is *w* – X*e*. If we alter X until the drop remains like Mahomet's coffin at rest under the force, $e=w/X$, which determines *e*. In practice it is better to measure the rate of fall under different electric forces, than to attempt to get gravity exactly balanced.

The value of *e* given by these methods is, within limits of experimental error, equal to the charge carried by an atom of hydrogen in electrolysis ; thus the fact that e/m for these charged particles is more than a thousand times that for the atom of hydrogen, must be due to an abnormally small value of *m* and not to an abnormally large one of *e*.

Number of Electrons in the Atom

Since the mass of an electron is only about 1/1800 of that of an atom of hydrogen, unless there are hundreds of electrons in the atom they will not account for more than a very small fraction of its mass ; it is therefore a matter of great interest to determine this number. This was first done at the Cavendish Laboratory in 1904 by measuring the scattering of X-rays when passing through a gas. X-rays behave like light waves of very small wave-length, and these on Maxwell's Theory are accompanied by electric and magnetic forces ; when these rays pass over an electron the electric forces will accelerate the electron. A charged body when accelerated gives out radiation, and the rate at which the energy of the radiation is emitted was shown by Larmor to be $2e^2f^2/3c$, where e is the electric charge on the moving body, f the acceleration, and c the velocity of light. If F is the electric force in the X-rays at the place of the charge, $f=Fe/m$, where m is the mass of the charged body ; the rate at which it is emitting radiant energy is thus $2e^4F^2/3m^2c$. The energy in unit volume of a beam of light when the electric force is F is, on the Electromagnetic Theory, $F^2/4\pi c^2$, and the energy flowing through unit area per second is $F^2/4\pi c$. The energy is emitted from the electron at the rate $8e^4\pi/3m^2$ times this, so that one electron per c.c. would scatter the fraction $8\pi e^4/3m^2$ of the energy of the X-rays passing through a cubic centimetre. If an atom contained n electrons, then, one atom would scatter n times this fraction, provided the distance between them were considerable when compared with the wave-length of the X-rays. If two electrons were only a small fraction of this wave-length apart, they would act like a double charge with a double mass, and we see from the formula, would scatter four times as much as a single electron. Again, the formula will not apply unless the frequency of the X-rays is large compared with the frequency with which the electron would vibrate if disturbed from its position of equilibrium in the atom.

344

When the preceding conditions are fulfilled, we see that when we have N atoms each containing n electrons in a c.c. of the gas, the energy in the scattered radiation will be $(8\pi e^4/3m^2)$ Nn times that in the incident X-ray radiation.

If we know the pressure of the gas we can, by Avogadro's law, determine N, and if we measure the fraction of the incident radiation scattered, we can determine n, the number of electrons in an atom. This was done by Barkla for several gases, and he found that for these the number of electrons in an atom was approximately half the atomic weight, except for hydrogen where there was one electron in the atom. This is in accordance with the result of modern investigations on the structure of the atom. This would make the part of the mass of an atom due to the electrons a very small fraction of the whole mass, but the fraction would be much the same for atoms of different chemical elements.

The expression for the scattering can be expressed very simply in terms of the radius of the electron. The mass m of an electron of radius a is given by the equation $m=2e^2/3a$; since the energy scattered by an electron is $8\pi e^4/3m^2$ times the energy of the radiation passing through unit area, it is 6π times the energy falling on the area of the cross-section of the electron.

On the quantum theory of light the scattering by electrons arises in a different way. When one of the photons in a beam of light strikes against an electron it is deflected, and produces light in the direction in which it is deflected ; it will also lose energy, and since the frequency of the light is proportional to the energy, the frequency of the light will be diminished, and the wave-length increased. The dynamics of the collision between photons and electrons have been worked out by E. H. Compton, who showed that it follows, from the principles of conservation of energy, that the increase in wave-length produced by a collision is independent of the frequency of the light and is 0·0243 Å ; this is too small to be appreciable for ordinary light, but for very hard X-rays the effect is very marked. When there is no appreciable loss of energy at a collision the amount of scattering is much the same on either theory.

Diffraction of Electrons: Electronic Waves

It was not until 1911, long after the discovery of X-rays, that it was proved that these rays could be diffracted ; there was strong evidence that they were light of very short wave-length, so short that it was infinitesimal in comparison with the distance between the rulings on an ordinary diffraction grating, so that these could not diffract the rays. In 1911, however, Laue had the brilliant idea that since a crystal consisted of a series of atoms arranged in regular order, it should act like a diffraction grating in which the intervals between the rulings were equal to the distance between neighbouring atoms, and would be able to produce diffraction effects for waves whose length was not very much less, or very much greater, than this distance. Using crystals as diffraction gratings, he obtained diffraction effects with X-rays, and was able to measure their wave-length.

Davisson and Germer in 1927 directed a beam of electrons at right angles to the face of a nickel crystal, and measured by an electroscope the number of electrons coming off in different directions. They found that the number of electrons in the scattered beam did not vary continuously with the angle of scattering, but that there were several directions in which the scattered electrons showed maxima of intensity, and that these directions corresponded to the direction of diffracted X-rays of a particular wave-length. This wave-length, λ, was in good agreement with an equation given by Prince Louis de Broglie :

$$\lambda = h/mv,$$

where h is Planck's constant, and m and v the mass and velocity of the electron

Two months after the publication of these results, my son, Professor G. P. Thomson, published photographs obtained by sending a beam of homogeneous cathode rays—*i.e.* a beam where all the electrons had the same velocity —through a very thin film of collodion. This showed a bright central spot due to electrons which had passed through without being deflected. Around this there were rings, and the diameter of these was in agreement with de Broglie's equation.

The chemical composition of collodion is too complicated and uncertain to allow of a test of the agreement of the experiment with theory, so Thomson tried very thin films of the various metals for which the dimensions of the lattices, formed by the atoms in the crystal, had been determined by experiment with X-rays. The first photograph published of the rings shown by thin gold is shown in Plate I (Fig. 1). He made a very large number of experiments and found numerical agreement between the results of his experiments and de Broglie's formula within the limit of experimental errors; actually the agreement was within a very few per cent. From this he drew the conclusion that the electron is accompanied by trains of waves, whose wave-length is inversely proportional to the velocity, and that these waves guide the electrons. That the various rings of the photographic plate are due to electrons and not to waves travelling independently, is shown by the fact that if after passing through the film, and before reaching the photographic plate, the electron is deflected by a magnet, the whole system—central spot and rings— is deflected, and the relative intensity of the various parts preserved.

As the result of his experiments, Thomson came to the conclusion that each electron is associated with a wave whose wave-length is approximately h/mv, the length of the train being at least 50 wave-lengths and the breadth of the wave front at least 30×10^{-8} cm. When the electron is moving with uniform velocity and is in a steady state, if there is any energy in the train of waves it must travel with the velocity of the electron, which is small compared with that of light.

The explanation of " beats " in sound waves shows that there may be two velocities associated with waves. Suppose that a disturbance represented by $a \cos (pt - mx)$ is superposed on another represented by $a \cos (p't - m'x)$, the resultant is $a \cos (pt - mx) + a \cos (p't - mx)$

$$= 2\, a \cos \tfrac{1}{2}((p - p')t - (m - m')x)$$
$$\cos \tfrac{1}{2}((p + p')t - (m + m')x)$$

or if p and p' are nearly equal

$$2\, a \cos \tfrac{1}{2}((p - p')t - (m - m')x) \cos (pt - mx).$$

This can be regarded as a vibration $\cos (pt - mx)$, whose amplitude varies as $a \cos \tfrac{1}{2}((p - p')t - (m - m')x)$. This represents a wave whose velocity is $(p - p')/(m - m')$ or in the limit dp/dm. Now the energy in the wave at any place depends upon its amplitude, so that dp/dm represents the velocity of propagation of the energy, while p/m represents the velocity of the phases of the wave $\cos (pt - mx)$.

The quantity measured in experiments on electron waves is the wave-length of the waves represented by cos $(pt - mx)$: why is this connected with the velocity of the electron ? It is because in certain important cases, *e.g.* that of the propagation of waves through a medium which has an intrinsic frequency of its own, the product of the phase velocity and the energy velocity is equal to the square of the velocity of light. If p_0 is this intrinsic frequency the equation of wave motion through the medium is

$$\left(\frac{d^2}{dt^2}+p_0{}^2\right)\phi=c^2\frac{d^2\phi}{x^2},$$

or if $\phi=\cos (pt - mx)$

$$p^2=c^2m^2+p_0{}^2.$$

The phase velocity is p/m, the energy velocity dp/dm, the product of the two is $\dfrac{pdp}{mdm}$ or $\dfrac{dp^2}{dm^2}$; this by the preceding equation is equal to c^2. From this equation too we get

$$\frac{\lambda u}{\sqrt{1-\dfrac{u^2}{c^2}}}=\frac{2\pi c^2}{p_0}=\text{constant.}$$

Here λ is the wave-length of the phase velocity $=2\pi/m$, and u the energy velocity which is that of the electron. If we suppose that $2\pi c^2/p_0=hm_0$, where m_0 is the mass of an electron at rest, and h Planck's constant, this is exactly the same as de Broglie's equation, with which Professor Thomson's results were in excellent agreement. But though there is formal agreement between the theoretical expression and the experiments, the numerical results give rise to difficulties. From the sharpness of the rings Professor Thomson concluded that there must be at least 50 wave-lengths in a train of waves, and that the wave-lengths were comparable with 10^{-8} cm. Thus the dimensions of the train would be enormous compared with the usual estimate, 10^{-13} cm. of the diameter of an electron. This is not quite conclusive, however, for the wave-length of visible light emitted by a luminous atom is also enormously greater than the linear dimensions of the atom, and the quantum of light must be a train of a large number of wave-lengths to explain the optical effects ; this long train must have been manufactured within the short atom and then expelled, and it is possible that the long train of electronic waves may have started at a distance from the electron very much less than the wave-length of the electronic waves.

Positive Rays

The cathode rays start from the cathode and move away from it. Goldstein, who devoted a long life to the study of the electric discharge through gases, and who made many discoveries of first-rate importance, found in 1886 that there were rays moving towards the cathode and striking against it. Using a cathode perforated by small holes he observed luminous pencils streaming through the holes. These he called " Kanalstrahlen " ; the colour of these pencils was not the same, however, as that of the cathode rays. A more fundamental difference was that these rays were not appreciably deflected by magnetic forces which produced large deflections of cathode rays ; indeed it was not until sixteen years after their discovery that their deflection by a magnet was observed by Wien, who found that it was in the opposite direction to that of cathode rays, so that these rays carried positive charges.

He obtained, too, direct evidence of these charges, and by measuring the deflection under electric and magnetic forces, he obtained 10^4 as the maximum value of e/m; this is the value for the hydrogen atom in electrolysis. I thought the study of these rays might throw light on the nature of the carriers of positive electricity. Are they, for example, all of one type, like the carriers of negative charges? The method I used for this purpose was to send a narrow pencil of rays through the space between two parallel brass plates and apply in the space between these plates electric and magnetic forces, both at right angles to the plates. If the plane of the paper is at right angles to the beam of electrified particles, and also to the plates, and if AB and CD are the lines of intersection of this plane and the plates, then, as the beam passes between the plates, the electrified particles will be acted upon by two forces at right angles to each other : a horizontal one due to the electric field and a vertical one due to the magnetic. The rays when they

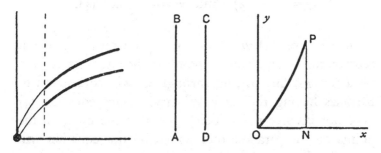

strike against a photographic plate produce a spot where they hit the plate. Suppose this plate is placed at right angles to the beam of rays, and that when neither electric nor magnetic forces are acting, the rays hit the photographic plate at O, and that when the electric and magnetic forces are on, O is displaced to P.

Then if Ox and Oy are respectively horizontal and vertical lines through O, and PN is vertical, ON is the displacement due to the electric force and PN that due to the magnetic force.

It can be shown that

$$\text{ON}=\frac{eX}{mv^2}\text{A}, \quad \text{PN}=\frac{eH}{mv}\text{A}$$

where e is the charge, m the mass and v the velocity of the particle, H the magnetic and X the electric force, and A a quantity depending only on the geometry of the system, *i.e.* on the length of the path of the particle between the plates AB and CD and on the distance from them of the photographic plate. Writing y for PN and x for ON we see

$$x=\frac{Xe}{mv^2}\text{A} \quad (1) \quad \text{and} \quad y=\frac{He}{mv}\text{A} \quad (2),$$

$$y=\frac{Hv}{X}x \quad (3) \quad \text{and} \quad y^2=\frac{e}{m}\frac{H^2A}{X}x \quad (4).$$

From equation (3) y/x does not depend on the mass but only on the velocity; all the points in the photograph lying on a straight line passing through O will correspond to particles having the same velocity. Since equation (4) does not involve the velocity but only the mass, all the points on the parabola represented by this equation will correspond to particles having the same mass. If there are particles of different mass in the beam of rays there will be a different parabola for each type of particle.

It follows from equation (1) that if T be the kinetic energy of an electrified particle, $x=XeA/2T$. The energy of the particles is due to the forces acting upon them before they passed through the cathode ; these, if the charges on the particles are equal, will be the same for particles of all kinds ; in particular the maximum energy which the different particles acquire will be the same, and therefore the minimum value of x. Thus the parabolas will all stop at the same distance from the vertical. If some particles had a double charge of electricity when in the discharge-tube, and lost the extra charge before passing through the parallel plates, they would have acquired twice the normal amount of energy, so that their parabolas would reach to half the normal distance from the vertical.

On trying this method I met at first with great diffi-culties. This was due to the fact that these positively charged particles have very little penetrating power com-pared with the cathode rays, so that unless the pressure is very low they will not reach the photographic plate. We are then met with the difficulty that the discharge required to produce these rays will not pass when the pressure is low enough to enable the positive rays to reach the plate, and if the pressure in the discharge-tube is higher than on the other side of the cathode, gas will flow through the holes and raise the pressure on that side. To stop this as much as possible a long straight tube of fine bore, like a hypodermic needle, was fixed in the hole so that the gas had to pass through this tube, which offered a considerable resistance and made the rate of leak slow, but still large enough to require continual pumping to enable the rays to reach the plate. At first all we got on the plate was a straight line passing through the origin. The value of e/m for the tip of this line was 10^4, and we got this line and no other whether the gas introduced into the discharge-tube was air or nitrogen or CO_2, and it looked as if the carriers of the positive electricity were all of the same kind and were atoms of hydrogen. However, by using very large vessels for the discharge-tube, which allowed the discharge to pass at a lower pressure than the smaller ones we had previously used, we got two lines on the plate and the value of e/m for the tip of the second line was $\frac{1}{2}10^4$, the value for a molecule of hydrogen. These two hydrogen lines were the only ones we got until we put some helium in the tube, when a third line with e/m, $\frac{1}{4}10^4$, the value corresponding to an atom of helium, appeared. This was very important, for it was the first evidence we got that the positive particles could be other than hydrogen atoms or molecules. At this stage, through the generosity of Mr T. C. Fitzpatrick, a plant for making liquid air was installed in the laboratory, and we were able to use Dewar's method of producing a vacuum by absorbing the gas by charcoal cooled with liquid air. Now the worst of our troubles were over, we had no difficulty in getting parabolas corresponding to the atoms and molecules of the different gases in the discharge-tube—nitrogen, oxygen, carbon. The reason we had only got hydrogen to begin with is that there is always hydrogen condensed on the glass walls of the discharge-tube which is given off when the discharge passes through it. These light atoms and molecules have much greater powers of penetration and can reach the plate when the atoms and molecules of the heavier gases are unable to do so. Helium, the next lightest gas, has also great penetrating power, and it was able to reach the plate almost as well as hydrogen. These experiments show that the positive charges, unlike the negative ones, can be carried by the atoms and molecules of all gases, elements or compounds, and suggest that a positively charged particle is just one that has lost an electron. The difference made by the improvement in the vacuum is illustrated in Plate I (Fig. 2), taken after we used liquid air. The gas in the tube was that left in after the gas had been pumped out until the pressure was very low. There are parabolas due to particles for which $10^4(m/e)$ equals 1, 2, 12, 14, 16, 28, 44, due to atoms and molecules of hydrogen, atoms of carbon, nitrogen and oxygen, molecules of nitrogen, carbon monoxide and carbon dioxide respectively. Some of these are due to gases coming from the walls of the tube. Each kind of charged carrier produces its own parabola on the plate ; there are as many parabolas as there are different kinds of carriers. We get a spectrum of the gas and from an inspection of the plate we can determine not only the number of kinds of carriers, but also from the dimensions of the parabolas the atomic or molecular weight of each carrier. This type of spectrum enables us to determine the nature of the gases inside the tube and thus provides a method of chemical analysis.

The positive ray spectrum has for this purpose many advantages over ordinary spectrum analysis. If a spectroscopist observes a line unknown to him in the spectrum of a discharge-tube, the most he can infer, without further examination, is that there is some unknown substance in the tube, and even this might be doubtful, as the new line might be due to some alterations in the conditions of the discharge. But if we observe a new parabola in the positive ray spectrum all we have to do is to measure the parabola, and this at once tells us what is the atomic weight of the particle which produced it. By giving long exposures we can make the method exceedingly delicate, and detect the presence of a trace of gas too small to be detected by spectroscopy. The amount of gas required is very small, as the pressure of the gas is exceedingly low, generally less than one-hundredth part of a millimetre of mercury.

Another very important advantage is that this method is not dependent upon the purity of the gas ; impurities merely appear as additional parabolas in the spectrum and do not produce any error in the determination of the atomic weight. The rays are registered on the photograph within much less than a millionth of a second after their formation, so that when chemical combination or decomposition is going on in the gases in the tube, the method may disclose the existence of intermediate forms which have only a transient existence, as well as that of the final product, and may thus enable us to get a clearer insight into the process of chemical combination.

I took a great many photographs of the parabolas when different gases were put in the discharge-tube. Among these were some samples of gases obtained from the residues of liquid air. This gave, along with other parabolas, a strong parabola of neon, atomic weight 20 ; and the neon parabola was always accompanied by one close to it, corresponding to a particle of atomic weight 22, and by another much fainter one, corresponding to a particle for which e/m was 11, the value it would have if the atom of the line 22 had a double charge of electricity. The atoms of nearly all the elements except hydrogen can carry a double charge, while it is very exceptional for a molecule to do so. This was against the line 22 being due to a hydride NeH_2. The same objection would apply to its being a molecule of CO_2 with a double charge, which would give the parabola with $m/e=22$, and in addition to this difficulty there is the fact that all the CO_2 can be removed from the gas without affecting the strength of the line. Whenever neon was present the line was there, and it never occurred except accompanied by the neon line. It was the first example of an Isotope. Two elements A and B are isotopes if their chemical properties are identical but their atomic weights different.

Professor Soddy shortly before the discovery of the isotope of neon had suggested the possibility of the existence of such things as isotopes. Since the chemical properties of an atom depend on the arrangement of the electrons near its surface, they are only skin deep. If from the centre of the atom a proton with a positive charge and an electron with a negative one were removed simultaneously, the electric field at the surface of the atom would not be changed and therefore the arrangement of the electrons at the surface not affected, so that the chemical properties would be unaltered while its atomic weight would be diminished by unity. Neither chemical nor spectroscopic tests could distinguish between them, while physical ones such as trying to separate them by diffusion have not been successful, so the positive ray test is the only one available.

Aston, who has made very extensive and valuable experiments on isotopes and has discovered isotopes for most of the elements, did not use the parabolic method, but one where rays of different velocity but of the same mass were brought to a focus and the effect on the photographic plate thereby increased. The electric and magnetic fields were placed in series, instead of being superposed as in the parabola method, and the magnetic force was at right angles and not parallel to the electric. This arrangement makes the force exerted by the magnetic field on the charged particles parallel to, but in the opposite direction to, that exerted by the electric. It focusses the rays of different velocity on the same principle as two prisms, made of different kinds of glass and arranged so as to bend rays of light in opposite directions, can bring rays of light of different colours to the same focus. The two prisms in the optical experiment are analogous to the electric and magnetic fields which bend the positive rays in opposite directions in the electric one. The variation of the amount of bending with the wave-length in the optical experiment is analogous to the variation of the bending with the velocity of the rays in the electric one. And, just as we cannot obtain achromatism by using two prisms of the same kind of glass, so we cannot focus the positive rays by using two electric or two magnetic fields while we can when one field is electric and the other magnetic. The reason of this is that the change of the deflection with the velocity, which corresponds to the dispersion of light in the optical problem, varies inversely as the square of the velocity in the electric field, and only inversely as the velocity itself in the magnetic. There is thus a difference in the law of dispersion in the two fields, just as there is in prisms made of different kinds of glass in the optical one.

Under good conditions the parabola method need not be inferior in sensitiveness to the focus method. Indeed, quite lately, Zeeman, using the parabolas, detected an isotope of argon which had escaped notice by the focus method. The parabola method has for many purposes considerable advantage over the other. It shows the whole spectrum at a glance : the parabolas have a structure of their own ; sometimes there are concentrations in particular spots ; sometimes besides the parabolas there are straight lines on the plates. The beam of rays sometimes contains negatively electrified particles as well as the positive ones ; these give rise to another set of parabolas. All these peculiarities, which throw much light on the processes taking place in the discharge-tube, are much more easily detected and investigated by the parabola method than by the other.

When the pressure is not too low to allow of collisions between the molecules of the gas in the tube with the charged particles on their way to the photographic plate there are, in addition to the parabolas, lines starting from the position of the undeflected spot. These lines are generally straight; they reach in some cases to one of the parabolas and then stop. In other cases they stop before reaching a parabola. The first kind of lines are due to particles some of which have lost their charges while going through the electric and magnetic fields and so have not experienced the full deflection. The lines which stop before they reach a parabola are due to particles which have lost their charge before passing out of the fields. Again, the parabolas themselves show differences which give important information. The majority of the parabolas start from points in the same vertical line if the magnetic deflection is in this direction. The horizontal distance from the line of no electrostatic deflection is inversely proportional to the energy of the charged particle, so that if the parabolas all start at the same distance from this line, the maximum energy possessed by the particles is the same for all. This we might expect as the charges carried by them are the same, and the maximum potential difference through which they fall, that between the electrodes in the tube in which they are produced, is the same. There are, however, some parabolas whose tips are nearer the vertical line than the others, showing that their particles possess an abnormal amount of energy; a case of this kind is shown in Plate I (Fig. 3), where the distance of the tip of one of the parabolas is only half that of the others, showing that the maximum energy of this kind of particle is twice the normal. This is what would happen if some of the particles lost two of their electrons instead of one before passing through the cathode; they would have a double charge, and so receive in the discharge-tube twice the normal energy, while if they regained an electron after passing the cathode and before reaching the electric and magnetic fields, they would lie along the parabola corresponding to the normal charge ; those which did not lose their second charge would appear along the parabola with a double value of e/m, and this second parabola can always be found along with the one showing the prolongation.

The discharge decomposes the gas through which it passes, splitting up the molecules of the elementary gases into atoms, and the molecules of compounds into any combination which can be made of the elements of which they are composed. The discharge is accompanied by association as well as decomposition : thus when marsh gas was in the tube, Conrad [1] got the positive ray spectrum shown in Plate I (Fig. 4). There are in this spectrum lines corresponding to

$$C, H_1, \quad CH_1, \quad CH_2, \quad CH_3, \quad CH_4$$
$$C_2, H_2, \quad C_2H, \quad C_2H_2, \quad C_2H_3, \quad C_2H_4$$
$$C_3, H_3, \quad C_3H, \quad C_3H_2, \quad C_3H_3, \quad C_4H_4$$

It seems that every possible combination of the atoms of the gas through which the gas passes are produced without any limitation by considerations of valency. It must be remembered that these combinations are recorded in a fraction of a millionth of a second after they are produced, and so may have only a very short life.

[1] *Phys. Zeits.* 31, p. 888, 1930.

PLATE I

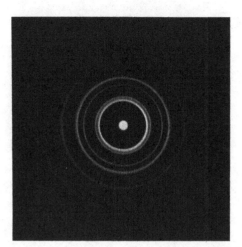

Fig. 1

Fig. 2

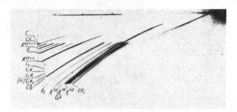

Fig. 3

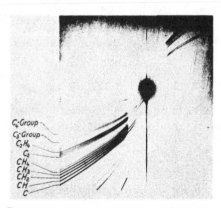

Fig. 4

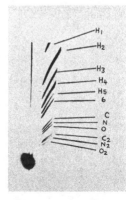

Fig. 5

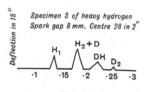

Fig. 6

Negatively charged Atoms

In the photographs of the parabolas of the positive rays we often find parabolas where both the electric and magnetic deflection of the particles are in the opposite direction to those of the normal parabolas. This shows that these parabolas are formed by particles which carry a negative charge. Before they passed through the cathode they were positively charged, and they owe their velocity to the action on them, when in this condition, of the electric force in the discharge-tube. After passing through the cathode they attract first one electron which neutralises them, and then another which gives them a negative charge. These negatively charged particles are, in comparison with the positively charged ones, more conspicuous at high pressures than at low. The negatively charged hydrogen atom can generally be detected down to fairly low pressure. At these pressures the negatively electrified molecule cannot be detected, while at higher pressures it is unmistakable. At still higher pressures I have seen the parabola due to negatively charged hydrogen atoms as bright as that due to the positive ones.

The electrochemical properties of the gases have a more conspicuous influence on the occurrence of these negative rays than on any other phenomenon connected with positive rays. For example, the atoms of the electronegative elements oxygen and chlorine are remarkable for the ease with which they acquire a negative charge. Carbon and hydrogen do so too, though they are not regarded as electronegative elements ; I have never seen the negative parabolas for the inert gases, helium, nitrogen, neon, argon, krypton, xenon, nor that of mercury, though the positive ones were very strong. Another effect which is very useful in interpreting positive ray photographs is that negatively electrified molecules, with the exception of those of hydrogen, oxygen, carbon, and these but rarely, are not found among the positive rays. Some radicles such as OH, CH_2, CH, occur with a negative charge when hydrocarbons are in the tube.

The prolongations of the parabola test whether a parabola is due to an atom or a molecule. All atoms except those of hydrogen seem under certain conditions to give prolonged parabolas while molecules very seldom do so. The parabolas due to the atoms of mercury have abnormally long prolongations—the one corresponding to the singly charged atom is prolonged until its tip is only one-fifth of the normal distance from the vertical, showing that some of the atoms have lost 5 electrons; there are parabolas corresponding to atoms with 2, 3, 4, 5 charges respectively. I have also detected atoms of nitrogen and oxygen with 3 charges.

Electric Method

The photographic method, however, is not metrical: the intensity of the parabolas for different elements is not proportional to the number of particles producing them. The atoms of the light elements produce a much greater photographic effect than those of the heavier ones. Thus the parabolas for H^+ and H_2^+ may be by far the strongest on the plate, when the amount of hydrogen in the tube is but a small fraction of that of other gases.

A method which is much more metrical is to use instead of the photographic plate a metal plate in which a parabolic slit is cut; behind the slit is a Faraday cylinder connected with an electrometer; the positively charged particles passing through the slit charge up the electrometer and the deflection of the electrometer measures the charge received in a given time. It is not necessary to have more than one slit even when there would be many parabolas on a photographic plate, for one parabola after another can be driven on to the slit by altering the magnetic field used to deflect the particles. The method is very definite; a change of a few per cent in the magnetic force is enough to make the difference between no deflection of the electrometer and a very large one. When the particles lose their charges before reaching the plate the two methods do not measure the same thing : the photographic method records all particles, whether charged or uncharged, which reach the plate, the electrometer method only those which are charged. Information about the stability of different kinds of charged particles can be got by applying both methods to the same gas. Thus at certain pressures the strength of the photograph of the parabola due to the negatively charged hydrogen atom is comparable with that of the positively charged one, but while the positive one gives a large deflection with the electrometer method, the negative one does not give one large enough to be measured, showing that the negative one does not retain its charge even for the short time, less than $1/100000$ of a second, required to pass from the place where it was formed to the electrometer.

Again, one of the first things discovered by the photographic method was the existence of H_3^+. Under suitable conditions its parabola may be of very considerable strength, stronger than many of the other parabolas, while when tested by the electrometer its effect is only a very minute fraction of that due to the lines it excelled before, showing that the greater number of the particles lose their positive charge, *i.e.* become neutral, in a minute fraction of a millionth of a second. The other form of hydrogen with the same molecular weight DH, when D is the atom of the isotope of hydrogen, whose mass is 2, gives an effect on the electrometer of the same order as that on the photographic plate, behaving in this as in other respects like H_2. Though H_3^+ is so evanescent, H_3 itself is much more durable, for when H_3 has once been observed in a tube, then, though the tube has not been sparked for several days, H_3 will at once appear when sparking is resumed though it gets fainter as time goes on. Presumably it adheres to the glass wall of the tube and is, like layers of H and CO, not easily detached.

Mass, Momentum and Energy of Moving Charges

On Maxwell's theory the velocity of light is equal to a well-known electrical constant whose value had been determined before the theory had been proposed. Other theories, for example the elastic solid theory, give expressions for it in terms of quantities we know nothing about, but his is the only one which says that the velocity must be 3×10^{10} cm./sec.

It may at first sight seem surprising that a velocity which appears in such a prosaic form as the ratio of two electrical units should be of such outstanding importance in modern physics, but we must remember that on modern views matter is supposed to have an electrical structure, and that on Maxwell's theory light is an electrical phenomenon.

Maxwell only applied his theory to a medium devoid of electric charges, but if we apply it to the case of a moving electric charge we arrive at conclusions of a very fundamental character, similar to some of those which were at a later date arrived at from considerations of relativity.

Let us take a very simple case to begin with, that of a charged sphere moving in a straight line with a velocity very small compared with that of light. Since the charge is moving, the electric force at a point in its neighbourhood will change, and on Maxwell's theory changes in elec-
tric forces produce magnetic
forces, hence there must be
magnetic force in the space

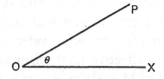

around the moving sphere. Maxwell's theory applied to this case shows that the moving charge will produce at a point P a magnetic force equal to $eu \sin \theta/r^2$, where e is the charge measured in electromagnetic unit, u the velocity of the charge, OX the direction in which it is moving, and θ the angle POX, O the centre of the charge and $r=$OP. The direction of the force is at right angles to the plane POX, $i.e.$ the plane which contains the radius vector OP and the direction of motion OX. But wherever there is magnetic force there is energy. The energy per unit volume at a place where the magnetic force is H is $H^2/8\pi$. In the case of the moving particle, $H=eu \sin \theta/r^2$, so that at P the energy per unit volume is $e^2u^2 \sin^2 \theta/8\pi r^4$; integrating this over the region outside the surface of the sphere, we find that the energy outside the sphere is $2e^2u^2/3a$, where a is the radius of the sphere.

If M is the mass of the sphere when it is without a charge, the kinetic energy of the charged sphere will be $(M+4e^2/3a)u^2/2$. Thus the effect of the charge has been to increase the mass by $4e^2/3a$; now $e^2c^2/2a$, c being the velocity of light, is the energy in the space outside the charged sphere, so that the charge on the sphere has increased its mass by $8/3c^2$ times the increase in the energy. The increase in mass is thus proportional to the energy of the electrified system. Though the charge of electricity may be the same, the increase in the mass will be greater when it is spread over a small sphere than over a large one. The increased mass is distributed throughout the region surrounding the charged body. We may compare this increase in mass with that which occurs when a sphere is moving through water. If M is the mass of the sphere, its effective mass when moving through water is not M but $M+m/2$, where m is the mass of an equal volume of water. The reason for the increase in mass in this case is obvious. The sphere cannot move without setting the surrounding water in motion, so that when it moves the mass in motion is not only the mass of the sphere, but also the mass of the water set in motion by the sphere.

The increase in mass is the same for positively as for negatively electrified particles if they are of the same size, and we can prove without difficulty that if a number of charged particles are separated by distances large compared with their diameters, as they are in an atom regarded as a collection of protons and electrons, the energy of the atom will be the sum of the energies of the individual protons and electrons of which it is composed. It has just been shown that the mass of each constituent is in a constant proportion, $8/3c^2$, to its energy. Therefore the sum of the masses of the constituents will bear the same proportion to the sum of their energies. Hence the mass of the atom will bear a constant proportion to its energy. An uncharged body may be regarded as one containing as many protons as electrons. The proportion between mass and energy would be again the same. Thus the energy of any body is equal to the product of its mass and a multiple of c^2 ; this result was arrived at before it was deduced from the principle of relativity.

We can, however, go further than this. We have supposed that the velocity of the particle was so small compared with that of light that the squares of the ratio of the two may be neglected. We can, however, get from Maxwell's theory the solution without this limitation. It follows from the theory that when an electrical charge is moving, the distribution of electrical force round the particle is not the same as when the particle is at rest. In the latter case the force is symmetrically distributed round the particle. If O is the centre of the particle, the force at P is e/OP^2 whatever may be the direction of OP, but when the particle is moving in the direction OX, the force depends on the angle between OP and OX. It has been shown by Lorentz and Heaviside that if x, y, z are the co-ordinates of a point on a line of electric force when the charge is at rest, x^1, y^1, z^1 the co-ordinates of the same point when it is moving with velocity u, then if $k^2 = (1 - u^2/c^2)$

$$x^1 = kx, \quad y^1 = y, \quad z^1 = z.$$

These equations show that the system of lines of force connected with the charge suffers a uniform contraction in the direction of x, *i.e.* in the direction in which the charge is moving, and thus the lines of force tend to set themselves at right angles to the direction in which they are moving. When the particle is moving with the velocity of light $u = c$, and $k = 0$, so that $x^1 = 0$, *i.e.* all the lines of force are at right angles to the direction of motion of the charge. If we regard the charged body as merely the terminus of the lines of force, as would for example be the case if the charge were a hole in the ether on which lines of force terminate, then, if when at rest they terminated on a sphere of radius a, represented by the equation $x^2+y^2+z^2=a^2$, they would when in motion terminate on the ellipsoid

$$\frac{x_1^2}{k^2}+y_1^2+z_1^2=a^2.$$

This is called the Heaviside Ellipsoid. The shape of the charged body would be changed by its motion.

The Mass of the Moving Charge

The momentum per unit volume at any point (x, y, z) in the electric field is equal to the vector product of the electric and magnetic forces. From this it follows that the momentum parallel to x is per unit volume

$$\frac{e^2u}{4\pi k^2}\frac{y^2+z^2}{\left(\dfrac{x^2}{k^2}+y^2+z^2\right)^3}$$

and integrating this throughout the region *outside the Heaviside Ellipsoid* whose boundary is $x_1^2/k^2+y_1^2+z_1^2=a^2$ we find that the momentum $=\dfrac{2}{3}\dfrac{e^2u}{ak}$.

Since the momentum is the product of the mass and the velocity, the mass of this moving charge is

$$\frac{2}{3}\frac{e^2}{ak}=\frac{2}{3}\frac{e^2}{a\sqrt{(1-u^2/c^2)}},$$

a result which was first obtained in a different way by H. A. Lorentz. Thus the mass of the moving electron varies as $1/\sqrt{1-u^2/e^2}$ and therefore increases as the velocity increases, and at the same rate as that which subsequently was indicated by the principles of relativity.

On this view the mass, momentum and energy of the charged sphere are distributed throughout the medium around it, and are not in the sphere itself. We shall call this medium, in which Maxwell's equations are assumed to hold, the ether, and if we assume the electrical theory of matter, *i.e.* that it is composed of electrons and protons, it follows that all mass, momentum and energy are in the medium surrounding the matter and not in the matter itself. If this is so, there must be some connecting link which binds a charged body to a portion of the ether, thereby endowing the body with mass, momentum and energy. These links I regard as supplied by the lines of force proceeding from charged bodies. We may use a hydrodynamical analogy to make this clearer, and compare lines of electric force with vortex filaments. When a long straight vortex filament moves through a liquid, the lines of flow of the liquid outside the filament are of two classes; there is a region extending to some way from the filament where the lines of flow are closed curves; the fluid in this region just moves round and round the filament and is carried along with it, and its mass may be regarded as the mass of the filament. Outside this region the lines of flow are not closed curves and the liquid does not accompany the filament. The volume of liquid moving with the filament depends on the " strength " of the filament and not on its volume, which may be quite insignificant in comparison with that of the liquid which it carries along with it. Again, the volume it carries with it is proportional to the square of the strength of the filament, so that two filaments close together will carry along with them twice as much liquid as if they were a considerable distance apart. We have seen that the effect of an increase in velocity of a charged body is to crowd the lines of electric force closer together and this, on the analogy between lines of electric force and vortex filaments, would produce an increase in its mass. On this view the ether outside the charged body is the seat of all mass, momentum and energy, and the dynamics of the electric field is the dynamics of this system and for this the mass, momentum and energy will remain constant, *i.e.* the Newtonian mechanics will apply. Part of this system is not connected up with matter; the Newtonian mechanics applies to the whole system, but not necessarily to a limited portion of it which can receive from or give up mass, momentum and energy to the other.

Let us take a very simple case to illustrate this point, one where the electric force is due to a charge of electricity at A and a magnetic pole of strength m at B. When there are both electric and magnetic forces at a point P there is momentum, and the momentum per unit volume is equal to the vector product of the electric and magnetic forces at P. From this we find by integration that the system has a moment of momentum about the line AB equal to *em*. Now suppose A and B are moving; if the motion is such that there is no velocity of A relative to B the direction of AB will not change and there will be no change in the moment of momentum in the ether. But suppose A is moving relatively to B; take the case when the relative velocity of B and A is at right angles to AB and equal to v, then in the time t^1 (supposed small) AB will turn through an angle vt^1/AB; the axis of the moment of momentum is turned through this angle, and the difference between the old value and the new is equal to a moment of momentum $(em/AB)vt^1$ whose axis is at right angles to AB and in the plane of the paper. There is thus a change in the moment of momentum of the medium. Next consider the moving charges: the moving charge e will exert on the magnet pole m a force emv/AB^2 at right angles to the plane of the paper, and the moving pole B will exert on the charge e a force of the same magnitude, but in the opposite direction; as the forces are equal and opposite there is no change in the momentum, but as they are not in the same straight line they will produce a couple whose moment is emv/AB, and whose axis is in the plane of the paper at right angles to AB. In the time t^1 this will produce a change in the moment of momentum of the pole and charge equal to $(em/AB)t^1$ which is the same in magnitude, but in the opposite direction to that produced in the ether, so that there is no change in that of the complete system.

CHAPTER XII

Physics in my Time

WHEN I began research in physics in the early seventies, the subject which was exciting most interest was the "Light Mill", or radiometer, invented by Sir William Crookes. This, in its commonest form, was four thin mica discs fastened to the ends of a horizontal cross which could spin round on a pivot supported in a glass cup. These were enclosed in a glass vessel which contained air at a very low pressure. The vanes were blackened with lamp-black on one side but were bright on the other, and when exposed to light they rotated round the pivot, the direction of rotation being the same as if the blackened side were repelled by the light. These light mills aroused the interest of the general public, and shopkeepers had them rotating in their windows to attract a crowd. The discovery of the radiometer by Crookes was a triumph of vigilance in observation and accurate measurement. When he was making very accurate weighings to determine the atomic weight of thallium, an element he had lately discovered, he found small discrepancies which he could not account for by any known source of error. He set himself to discover the origin of these, and by ingenious experiments convinced himself that they were due to the light falling on the balance. This led him to make *ad hoc* experiments on the effect of light on delicately poised pieces of metal, and hence to the discovery of the radiometer. At first the rotation was ascribed to the pressure of the light itself. Maxwell had shown that, on his electromagnetic theory of light, light ought to exert a pressure on surfaces on which it falls, and this pressure was detected in 1901, nearly twenty years after the discovery of the radiometer, by the Russian physicist Lebedew. This pressure is, however, far too small to account for rotations as great as those observed in the radiometer, and the rotation would be in the opposite direction. For suppose a light particle, a photon, fell on the blackened surface of a vane of the radiometer, it would be absorbed and would give up its momentum to the vane. If it were to fall on the bright surface, it would be reflected and would possess a momentum, equal in magnitude to that it possessed before striking the vane, but in the opposite direction, the reflecting vane would recoil with a momentum equal and opposite to that given to the reflected photon ; thus the reflection from the vane would double the momentum it would receive if the photon were merely absorbed. Thus the bright surface would be repelled by the light more than the blackened one, while in the radiometer the repulsion is less. Another very ingenious test was suggested by Osborne Reynolds, and carried out by Schuster. If the rotation were due to the pressure of the light on the vanes, it would not set up any rotation in the glass vessel in which the vanes were contained, but if it were due to effects inside the radiometer, then, by the principle that action and reaction are equal and opposite, if the vanes rotated in one direction the glass case must rotate in the opposite. The experiment was made in the laboratory of the Owens College ; the radiometer was suspended by a fine thread so that it could rotate freely if it wanted to. I can still remember the excitement and anxiety with which I waited for the verdict, and the relief on hearing that the case had rotated in the opposite direction to the vanes. The cause of the rotation of the vanes is that the blackened surfaces get hotter than the bright ones owing to their absorbing more light. When the molecules of the gas strike the blackened side they get hotter, and shoot off with a higher velocity than after striking the colder bright one, thus the kick or recoil is greater on the blackened surface than on the bright. His experiments on the radiometer led Crookes to study other phenomena in gases at very low pressures. In his time the production of a good vacuum was lengthy, tedious and arduous. The only means available were Sprengel or Töpler pumps ; both these required the continual raising and lowering of a vessel full of mercury, and it often needed a morning's pumping to get a vacuum good enough for the most important experiments. With the pumps now available a much better vacuum can be got in a few seconds. The introduction of electric light made the production of vacua a matter of commercial importance, as the air had to be pumped out of the lamps. This set inventors at work to produce a pump which should be automatic and rapid. The first satisfactory automatic pump was invented by Gaede in 1905, and from that time there has been a steady stream of automatic pumps working on a great variety of principles. Gaede himself has invented at least three different types of pumps, while successful pumps have been introduced by Langmuir, Waran and others. Some of these, it is claimed, can deliver five or more litres per second, and can reach a vacuum where the pressure is only one-thousand-millionth part of atmospheric pressure. This means that, of the molecules present in the vessel before the pumping began, only one in a thousand millions is left after it has been completed. And what is even more important for the investigation of the properties of matter is that the odds against one molecule coming into collision with another when it passes over a length of one metre is about 70 to 1. Thus, even in very large vessels, a molecule behaves as if no other molecules were present. At these pressures we can study the properties of individual molecules when they can do as they wish, whereas at considerably higher pressures we can only study their behaviour in a dense crowd when they have to do what the crowd dictates. The improvement in recent years in the production of high vacua is an example of the advantages which accrue to the study of any branch of science when it has industrial applications ; then the improvement of any technique becomes of commercial importance on which large sums may be profitably spent. The manufacture of vacua is an important industry. Kaye in his book *High Vacua* quotes Dr. Whitney as saying, when he was President of the General Electric Company of America, that the American public alone purchases " over a million dollars' worth of glass vacua yearly ". This was ten years ago, and in the interval the demand for vacua for electric lighting and wireless has much increased all over the world. I should think it would be an underestimate to put the trade in vacua, taking all countries into account, as £100,000,000 sterling a year.

The improvement in the pumps since 1905 has been in the speed with which they work. The method of producing vacua (absorbing the gas by charcoal cooled by liquid air) introduced by Dewar at the beginning of this century produced as good, and in some respects better, vacua than any method since introduced ; the only objection is that it is much slower. Owing to Dewar's discovery, physicists since the beginning of this century have been in possession of a means of studying matter in so rarefied a state that each particle acts independently of its neighbours. This has been of supreme importance to the progress of science. Crookes' experiments on the passage of electricity through gases were made before Dewar vacua had been invented, but by improving the Sprengel pump he was able to get down to pressures of about $\frac{1}{1000}$ of a millimetre, when the free path would be comparable with 7 cm., which would be low enough to give a molecule a fair chance of getting across the tubes he was using without a collision. His most striking experiments were made on cathode rays, a name given by the German physicist, Goldstein (who spent a long life in studying the electrical properties of gases), to something travelling in straight lines from the cathode when an electric current passed through a gas at low pressure, and producing phosphorescence when they strike against the walls of the box. They were first observed by Plücker in 1859. The nature of these rays became almost an international question. The German physicists, with the significant exception of von Helmholtz, regarded them as waves ; the English, I think without exception, thought they were atoms or molecules of the gas charged with negative electricity. The contest between these views went on with great vigour. Further research has shown that neither side was wholly right nor wholly wrong. The cathode rays are particles charged with negative electricity, but these particles are not atoms or molecules, but electrons —very much smaller bodies—and these electrons are accompanied by waves.

Crookes was a very skilful experimenter and he had the gift of arranging his experiments so that, in addition to their scientific importance, they were among the most striking and beautiful spectacles known in physics. With these he delighted, almost yearly, crowded audiences at the Friday Evening Discourses at the Royal Institution. His experiments led him to the conception of a fourth state of matter, a state in which the molecules were so far apart that they could travel across the vessel in which they were contained without coming into collision with other molecules. He said, and said quite truly, that when this is the case the properties of the gas will be quite different from those of a gas at higher pressures, when the molecules can hardly stir without hitting against other molecules. This is a necessary and very interesting consequence of any molecular theory of gases, and it was never more beautifully illustrated than by Crookes' experiments. It does not, however, tell us anything about the molecules. It was discovered later that the cathode rays were not molecules but very small particles called electrons, which had been knocked out of the molecules of the gas, and that these small particles were of the same kind from whatever kind of molecules they proceeded, showing that the electron is a constituent of every kind of substance. Crookes himself became a convert to the electron theory and said, " What was puzzling on the radiant matter theory is now precise and luminous on the electron ".

Crookes was in many ways unlike other English men of science. His well-waxed moustache gave him a somewhat foreign appearance ; he was not, as most of them were, connected with any university or college. He was a director of public companies, the proprietor and editor of a journal, gas inspector and expert witness, and at one time he owned a gold mine ; he did his work in a laboratory of his own. His training, too, had been different from that of his contemporaries. He left school when he was fifteen and went to the Royal College of Chemistry, where Hofmann, an eminent organic chemist, was Professor. Here he specialised in chemistry, and showed the individuality which was characteristic of him by working at inorganic chemistry, while the other pupils, including Perkins, the discoverer of aniline dyes, followed the example of the Professor, and worked at organic. His knowledge of mathematics was of the most rudimentary description, and he had never been through any course of instruction in physics. He picked up his physics as he wanted it for the research in which he was engaged, and could rely on being able to get Stokes' advice when he got into difficulties on theoretical questions. Stokes took great interest in his experiments and had a very high opinion of them; he says (*Memoir of Sir George Gabriel Stokes*, by Sir Joseph Larmor, vol. i. p. 353), "For enlarging our conceptions of the ultimate working of matter, I know nothing like what Crookes has been doing for some years". In the second volume of the *Memoir* some of Stokes' letters to Crookes are printed. They cover 133 pages; on one occasion at least there were two on one day. Crookes found Stokes' handwriting so difficult to read that he had to call to his aid the head printer of the *Chemical News*, a journal which Crookes had founded in 1861, and of which he was then proprietor and editor.

Stokes, it may be said, was one of the first to use a typewriter, but these were not available in time for the earlier letters. I remember hearing Stokes urging J. C. Adams to get one, and Adams said, "Why should I? People can read *my* writing."

Crookes' success was due not only to his skill as an experimenter, but also to his powers of observation. He was very quick to observe anything abnormal and set to work to get some explanation. He tried one thing after another in the hope of increasing the effect, so as to make it easy to observe and measure ; his work on the radiometer and the cathode rays are striking examples of this. In his investigations he was like an explorer in an unknown country, examining everything that seemed of interest, rather than a traveller wishing to reach some particular place, and regarding the intervening country as something to be rushed through as quickly as possible.

Between his investigations on thallium and those on the radiometer, *i.e.* between 1870 and 1874, Crookes was mainly occupied with investigations on what he called " psychic force ", which excited the interest of the public, and much criticism and obloquy from the great majority of men of science, who maintained that the results were worthless, and were produced by fraudulence on the part of the mediums. Now after sixty years, in which there have been many investigations made on the subject, opinion is much the same. Sir Oliver Lodge maintains that though a medium may be detected in fraud it does not necessarily follow that he has no psychic power. His view is that the medium is out to produce certain results, and that he will produce them with the least expenditure of effort, and, if the control is so lax that it is easier to produce them by conjuring than by his psychic powers, he will conjure.

The phenomena observed by Crookes were produced by mediums all of whom at one time or another had been detected, or had confessed to using fraudulent methods to obtain their results. By far the most important of these was D. D. Home, an American who came to London in

1869, and gave spiritualistic séances which became very fashionable and were very largely attended. He must have had very attractive manners ; he won the affection and confidence of Crookes and his wife ; he was adopted as a son by a wealthy lady who is said to have lavished £50,000 upon him (which she afterwards tried to get back) ; he was the Mr Sludge in Browning's poem, " Mr Sludge the Medium ". He was a medium of quite exceptional power, and had the great merit of being able to produce his effects in the light. This, of course, made trickery much easier to detect. Some conjurors, however, can do their tricks in a good light by distracting the attention of the observers. When dealing with mediums whose honesty is not beyond question it is necessary, before we can attach importance to their results, to be sure that no such distractions have taken place. Some of the most striking of Home's results could have been produced by a movement of a finger through a few inches. The possibility of hallucination cannot, I think, be regarded as altogether impossible. On page 265 of the *Life of Lord Rayleigh*, by his son, there is the following passage : " Lady Crookes . . . mentioned that at their séances they had been sitting round a particular table for some days, then they put it aside and took another. The first table came out from its corner apparently to attack the other, which leaped on to the sofa, was pursued by the first and they had a fight there. Crookes, when appealed to, said he knew nothing about motives, but corroborated the facts."

The phenomena obtained by Home were physical ones, tilting of tables, playing on accordions, production of floating luminosities and the like. Crookes' view was that these were not produced by spirits but by something coming from the medium himself, which he called psychic force. This force he supposed to act much as ectoplasm does in some modern theories.

The experiments which he describes were made in his dining-room, which was lit by a gas-jet of about 5 candle-power. The simplest and most important experiment was that supposed to prove that Home could alter the weight of a plank. A plank about 10 lbs. in weight was supported in a horizontal position : one end was attached to a spring balance to register its weight ; the other end rested on a knife-edge, which stood on a table, and in the experiments Home was sup-posed to press on the part of the plank between its free extremity and the knife-edge. Crookes and his wife sat on either side of him to see that the pressure was applied at this place, and with one foot on the foot of Home nearest to them. Under these conditions the spring balance indicated an increase of weight amounting in one case to nearly 12 oz. Crookes in June 1871 com-municated to the Royal Society a paper containing an account of this and some other experiments made with Home. The Committee of Papers, at their first meeting after the receipt of the paper, postponed their decision for a fortnight. Crookes heard of this : he wrote at once to Home : " I want you therefore to help me by giving me three evenings, between now and the 27th inst., at which I can repeat the increase of weight experiment in the presence of one or two other good witnesses, and then send in on the 28th an overwhelming mass of evidence which the committee can't reject". He sent in a further paper on June 28th, but it does not seem to have contained an account of any further séances with Home. The paper came before the Council of the Society and was unani-mously rejected. This course seems to me to have been entirely justifiable. Here was a discovery of stupendous importance, and at first sight of the utmost improbability, while the value of the evidence depended on the good faith of Home, a man of whose honesty there were grave suspicions, though Crookes believed in him implicitly. It is true that the experiments were not made in darkness and that some precautions were taken against fraud, but were these such as to make fraud impossible ? Might not Home sometimes succeed in distracting the attention of the observers and then do his trick ? He ran very little risk, for if he could at a séance not manage to elude the vigilance of the others, he had only to refrain from doing anything and say that his psychic power had failed. Crookes in a letter to Huggins says : " Home was in wonderful form last night, but he is the most uncertain of mediums, and it is quite as likely that the next time absolutely nothing will take place ".

Crookes published what were substantially the papers he sent to the Royal Society, in the *Quarterly Journal of Science* and also in the *Chemical News*. These met with severe criticism, which Crookes answered with much vigour ; he proved to be an excellent controversialist. His account of an interview he had with one of his most persistent critics, Dr. W. B. Carpenter, who was rather long-winded and had no inferiority complex, is very good reading. It is given in Fournier d'Albe, *Life of Crookes*, p. 213.

Crookes' active work on spiritualism was confined to the years 1870–74, when he returned to physics and discovered the radiometer. He retained, however, his belief in it until the end of his life and tried, after the death of his wife, to get into communication with her through a medium. He avowed his belief on many occasions; *e.g.*

SIR WILLIAM CROOKES

in his Presidential Address to the British Association at Bristol in 1898 he said : " I have nothing to retract ; I adhere to my already published statements. Indeed I might add much thereto." It is more than sixty years since Crookes gave up psychical researches, and we do not seem any nearer a conclusive proof of their reality than we were then. The aid of infra-red light has recently been used to determine what is going on in dark séances, but the results obtained so far are inconclusive. If we could get infra-red light intense enough to produce photographs with very short exposure, we might "film" a dark séance and detect any cheating in that way. As far as I know, no undisputed record of any spiritualistic phenomenon has been obtained. I have quoted Crookes' allusion to spiritualism in his Presidential Address : the subject turned up again at the same meeting, in a much less formal manner, at the Red Lion dinner. This is a function which occurs near the end of the meeting, whose aim is to get amusement rather than instruction out of the proceedings of the Association. Each President has a shield, with his name, crest and motto displayed upon it, hung on the walls of the reception room. The motto on Crookes' shield was " Ubi Crux, ibi Lux ". At the Red Lion dinner a distinguished chemist got up and unrolled somewhat nervously a shield which was a copy of Crookes' shield in everything except the motto, which was changed to " Ubi Crookes, ibi Spooks ". Crookes took this in very good part.

The interest taken in the subject of the passage of electricity through gases began to increase rapidly towards the end of the seventies. This was to a large extent because its vital importance in connection with the structure of molecules and atoms was beginning to be realised, and also because the phenomena associated with it were very various and of exceptional beauty. In England, besides Crookes, De la Rue and Hugo Müller, Spottiswoode and Moulton, and in Germany, Goldstein and Hittorf, made important advances at this early stage. Both De la Rue and Spottiswoode were of a class which has made more important contributions to science in this country than it has in any other. I mean the class who pursue science as a hobby and not as a profession, who work in laboratories of their own, and whose experiments are made at their own cost. Joule, who measured the mechanical equivalent of heat, was a brewer; Darwin a country gentleman. This class is not so numerous as it was, as the costs of a physical laboratory are so great that only very rich men can afford them.

De la Rue was the head of a great firm of manufacturing stationers, known all over the land since their name appeared on most of the playing-cards used in this country, and Spottiswoode was the head of a very large printing business. The experiments of De la Rue and Müller on the very interesting and beautiful phenomenon known as the " striated discharge " are of exceptional interest, and the plates they published in the *Philosophical Transactions* are still the best pictures we possess of this phenomenon, and have been reproduced in most textbooks dealing with electric discharge. This is probably because they used, for producing the discharge, a battery of a very large number of chloride of silver cells, which give a steadier current than induction coils, and, as the striations are steadier, the photographs have better definition. The Friday Evening Discourse at the Royal Institution at which De la Rue gave an account of these experiments was on a heroic scale. The preparation of the experiments for the lecture took, I believe, nine months ; a battery of 18,000 cells was set up in the Institution, and it was rumoured that many hundreds of pounds had been spent on the preparation. I have no difficulty in believing it. The weather unfortunately was wretched.

The two papers by Spottiswoode and Moulton on " Intermittence in the Electric Discharge " (*Phil. Trans.*, Part I, 1879, Part II, 1880) have never received the attention which I think they deserved. This may be because they are both very long and rather soporific. In spite of this they contain much that is valuable. It is interesting to note that, though Moulton was a Senior Wrangler and Spottiswoode President of the London Mathematical Society, there is not in these two long papers a mathematical symbol from beginning to end. The experiments described in these papers, and which gained for him the Fellowship of the Royal Society, were, as far as I know, the only experiments Moulton ever made. He had made none before either at school or Cambridge. They were made soon after he went down from Cambridge and whilst he was preparing for the Bar. After he had been called he rapidly acquired a very large practice, and had no time for research. His mind worked with wonderful rapidity ; the marks he got in the Mathematical Tripos when he was Senior Wrangler were more than double those of the Second Wrangler, who became a very eminent mathematician. I witnessed a striking instance of his mental agility, as I was in court during the trial of a suit which the owners of the patent for the " three wires sytem " of electric wiring brought against various electric lighting companies for alleged infringement of this patent. The patent had been taken out by Dr. John Hopkinson, but he did not at the time attach much importance to it, and sold it for a very moderate sum. It turned out, however, to be exceedingly valuable and many millions were involved in the trial. The Bar on either side was about three deep in Q.C.'s (it was in Queen Victoria's time), and Moulton was leading for the infringers. The plaintiffs called Lord Kelvin (then Sir William Thomson) as a witness on their side. After his examination-in-chief Moulton got up to cross-examine. He really had a very poor case, and at first seemed to be asking questions without any connection with each other. In answering one of these, a quite simple one, Thomson made an unaccountable slip and Moulton sprang at him in a flash : " Sir William, you say this". Thomson had only said it a few seconds before, so he said he had. "Then", said Moulton, " this follows " ; and it certainly did. From this admission he got another as a logical sequence, and so on, until these together seriously damaged the plaintiffs' case. Now he could not have foreseen that Thomson would make this slip, yet when it came he saw in a moment how to turn it to account. It was very interesting to watch how he went on from one step to the next, never letting Thomson get in sight of his first answer again. But this was not the end : when Thomson came out of the box some of his friends told him what he had done. He was very much perturbed and demanded to be recalled, and somehow or another he managed to be so. When he got into the box again, he lectured the Bench and the Bar for half an hour on the elementary principles of electricity, and nobody could get a word in. Counsel on the other side were jumping up every other minute saying, " My Lord, what has this to do with the case ? " " I don't know ! I don't know ! " said the Judge, and Thomson went on.

The importance of the work done by Lord Moulton during the war in organising the supply of high explosives for our troops can hardly be exaggerated. As soon as the war began, it was realised that the use of high explosives was to be on an altogether different scale from that of any previous war. Lord Moulton spent four years of uninterrupted work in providing our troops with an adequate supply. New workshops had to be built, others which had been used for other purposes adapted : the provision of the raw materials for the explosives was a matter of vital importance and great difficulty. Lord Moulton's efforts were so successful that the output, which at the beginning of the war was only one ton a day, rose to a thousand tons before the end.

Electrolytic Dissociation

In 1887 views about the nature of solution were put forward by Arrhenius and Van't Hoff which excited great interest and led to much discussion. They arose from experiments made by the botanist Pfeffer on what is known as " osmotic pressure ". There are membranes such as bladder or copper ferrocyanide which act like filters : they are permeable by water but not by substances dissolved in it. If such a membrane is placed between a solution and pure water, water will flow through the membrane from the water into the solution. To stop this flow a definite pressure must be applied to the solution ; this pressure is called the Osmotic Pressure of the solution. Pfeffer's experiments attracted the attention of Van't Hoff, one of the most brilliant figures in the history of chemistry. In 1874, when he was only twenty-two, he published a paper which was translated into French a year afterwards, with the title "La Chimie dans l'Espace", which may fairly be said to have been the origin of a large part of the work done in organic chemistry since its publication. He studied Pfeffer's measurement of the osmotic pressure of sugar solutions of various strengths and showed that, within the errors of experiment, "the osmotic pressure is equal to the pressure which a gas would exert if the number of its molecules per c.c. were equal to the number of molecules of sugar in the same space". He went further. Pfeffer had made measurements of osmotic pressure at different temperatures, and Van't Hoff showed that they followed roughly Gay-Lussac's law for the variation of pressure with the temperature of a gas when the volume is constant. This led him to the conclusion that in solution the solute is in the form of single molecules, and that these exert the same pressure as the same number of gaseous particles in a volume equal to that of the solution.

The direct measurement of osmotic pressure is very difficult except for very dilute solutions. For a solution of 1 gram equivalent per litre it is about 22 atmospheres, and semi-permeable membranes which would stand such a pressure are very difficult to make. Fortunately, however, it follows from thermodynamical principles that the osmotic pressure as defined above is proportional to the difference between the freezing point of the solution and that of water, and that this result is independent of any particular theory of how the pressure is produced, so that we can find the osmotic pressure by measuring the lowering of the freezing point. It also follows from thermodynamics that the osmotic pressure for dilute solutions, when the distance between the particles of the solute is so large that they do not exert any appreciable action on each other, depends only upon the number of the particles and not upon their kind, or whether they are all of one kind or are a mixture of different kinds. Using the freezing-point method, Raoult determined the number of *particles* in 1 c.c. of the solution. For weak solutions of sugar he found that this was equal to the number of *molecules* of sugar. He found, however, that this did not hold for solutions of salts and acids which are electrolytes. The osmotic pressure for these was always greater than that corresponding to the number of molecules of the salt or acid, and for very dilute solutions the results indicated that it was twice as great. The explanation of this increase given by Arrhenius was one of the greatest advances ever made in physical chemistry, and was as simple as it was daring. The experiments showed that for electrolytes the number of particles in the solution was greater than the number of molecules put in, hence some of the molecules must have broken up. Arrhenius had noticed that the electrical conductivity of the solution was proportional to the excess of the number of particles above the normal. This led him to the idea that when a molecule of a salt splits up in a solution it splits into two particles, one of which is positively, and the other negatively, electrified, *i.e.* into positive and negative ions. And that in very dilute solutions where there are twice as many particles as there are molecules there are nothing but charged ions and no molecules. That, for example, in a dilute solution of KCl there is no KCl, but only positively charged potassium ions and negatively electrified chlorine. This was a very startling result, for potassium itself is so violently acted upon by water that a piece of the metal thrown on water bursts into flame, and to suppose that an atom of it would not be acted upon in the water seemed as reasonable as to suppose that a man could escape getting wet by diving into the sea. The atom of potassium in the solution is not like an atom in the metal ; it is an atom carrying a charge of positive electricity. In the light of the electrical theory of matter, we regard the violence of the action between potassium and water as due to the effort of the potassium atom to get a charge of positive electricity ; when it has got this it is satisfied and no longer acts upon the water. On the theory of electrolytic dissociation, the potassium atom has got its positive charge and has no inducement to attack the water. At the time when Arrhenius published his theory, the effect of an electric charge on the properties of an atom was not understood. He must have been aware of the difficulties in the way of his theory and it was very courageous of him to publish the conclusions from which his experiments seemed to offer no escape.

It was quite natural that these difficulties should have prevented any general acceptance of the theory, and it met with very severe criticism. One prominent English chemist wrote, " Let us not pervert the morals of our students by talking glibly about atomic dissociation ". But though most chemists at first opposed the theory, it was supported from the beginning by Ostwald, a great German chemist, and the general acceptance of the theory was largely due to his clear presentation of it, and the experiments he made in its support. It is said that after reading Arrhenius' paper he went to Stockholm to discuss it with him. During the discussion there was a succession of taps on the door and Arrhenius went out and returned almost immediately. After the interview Ostwald asked what was the reason of this, and Arrhenius said that some

of his friends had heard of the visit, and, knowing what a
" leg-up " it would give the theory, had come to con-
gratulate him in the usual Swedish way by drinking a
glass of Swedish punch with him. Swedish punch is a
very potent drink, but it would have taken more than
that to muddle Arrhenius.

In dilute solutions the charged particles are so far from
one another that the forces between them do not produce
appreciable effects. In strong solutions, however, they
will influence the arrangements of the ions throughout
the solution; a study of this arrangement is of profound
interest and importance, and has engaged, and is engag-
ing, the attention of many eminent physicists.

Hertz and Electric Waves

In 1887 Heinrich Hertz, a young German physicist and
a pupil of von Helmholtz, demonstrated for the first time
the existence of waves of electric force. This discovery
was of transcendental importance both for pure science
and for its application to the service of man. It aroused
great interest all over the world, and in no place more than
in the Cavendish Laboratory, for the existence of electrical
waves had been suggested and their theory developed by
Clerk Maxwell, the first Cavendish Professor. Maxwell,
soon after taking his degree, had been very much interested
in Faraday's discoveries, and impressed by the extent to
which he had been guided in making them by his conception of lines of electric and magnetic force. Faraday
regarded an electric charge or a magnetic pole as the terminus of lines of electric force and magnetic force, stretching through space from charge to charge or from pole

to pole. On Faraday's view these lines of force were the origin of electric and magnetic forces. To him they were not merely geometrical lines, they had physical properties ; they were in a state of tension like stretched strings and thus produced the attractions between oppositely electrified bodies or between poles of opposite signs. Maxwell's first paper on electricity on " Faraday's Lines of Force" was published in 1855, and was mainly a translation of Faraday's views into mathematical language. In it he showed that these led to exactly the same values for forces between electric charges, and between magnetic poles, as the old theory of action at a distance. The next contribution of Maxwell to electrical theory was given in papers in the *Philosophical Magazine*, 1861 and 1862. In these papers he devises a model to illustrate Faraday's discovery of electromagnetic induction, *i.e.* that *changes* in magnetic force give rise to electric force. No one ever appreciated more than Maxwell the advantages gained in concentration of thought, and in the suggestion of new ideas, by considering a concrete case like a model, instead of relying upon algebraical symbols. He says : " For the sake of persons of different types of mind scientific truth should be presented in different forms, and should be regarded as equally scientific, whether it appears in the robust form and colouring of a physical illustration, or in the tenuity and paleness of a symbolical expression ".

His model was designed to illustrate the production of electric forces by changes in magnetic force. When he came to use the model he found that in it *changes* in electric force produce magnetic force. The introduction and development of this idea was Maxwell's greatest contribution to physics. He showed that it implied that the components of the electric force, as well as those of the magnetic, all satisfied the wave equations, and that the velocity of the waves was equal to a well-known quantity in electrical science called the ratio of the units. It is the ratio of the measure of a charge of electricity on the electrostatic system of units to the measure of the same charge on the electromagnetic system. This ratio had been measured by Kohlrausch and Weber some time before and found to be $3 \cdot 1 \times 10^{10}$ cm./sec., which agreed within the limit of errors of experiment with Fizeau's determination of the velocity of light. On the theories current before Maxwell's, electric waves could not exist, while on his theory every change in electric or magnetic force sent waves spreading through space with the velocity of light. Thus the velocity of light is the same as that of electric waves. This at once suggested that light waves were waves of electric force, and that their velocity could be calculated from purely electrical data. This is the outstanding feature of Maxwell's theory. It is the only theory which tells us that the velocity of light is 3×10^{10} cm./sec. Other theories such as the elastic solid theory give expressions for it in terms of quantities we do not know, but Maxwell's is the only one which tells us what the velocity of light is.

The papers in the *Philosophical Magazine* are a most fascinating account of the birth of a theory. I was so charmed by them when a boy that I copied them out in long-hand. Maxwell returns to his theory in a paper, "The Dynamics of the Electromagnetic Field", published in 1865. In this the model used in the earlier paper is not introduced. It had done its work by suggesting the existence of these new effects. In the later paper he postulates the existence of these effects and develops their consequences. The final presentation of his theory is given in his *Treatise on Electricity and Magnetism* published in 1873.

393

The part of his theory that attracted most attention was that light waves were waves of electric and magnetic forces. These waves, however, if Maxwell's theory is true, are only a small fraction of the electrical waves which, though we cannot see them, must always be passing through the space around us. The detection of these was vital for the establishment of the theory. It was many years after Maxwell's death before a satisfactory method of detecting these was available. We had to confine ourselves to testing whether light waves, in other respects than velocity, behaved in accordance with Maxwell's theory. Though the results of these were on the whole favourable, they could hardly be considered as conclusive, and could not be expected to convert the older men, who had been using for years other theories which had led them to great discoveries, and which were not inconsistent with any known electrical phenomena. It was quite different in the case of the younger men who were beginning their study of electricity and had no theories to give up. To them the beauty of Maxwell's theory made a strong appeal; " beauty is truth ", and there was no evidence that the theory was not true. Lord Kelvin's discovery of the oscillatory discharge of a Leyden jar, made long before, gave the means of producing electrical waves if Maxwell's theory were true. The difficulty was that the electric forces in these waves would be changing from one direction to the opposite millions of times per second, and no instruments were known which could make any response to forces of this type. These electric forces in the waves may be of considerable magnitude and could be detected with great ease if they kept acting in one direction for an appreciable time. Now the spark which passes between two metal balls placed close together, though it requires a considerable electric force to produce it, is, when such a force is available, produced in a very short time ; under favourable circumstances in much less than a millionth of a second. If the electric force in the wave acts in one direction for longer than the spark lasts, it can produce its effect before a force in the opposite direction comes to interfere with it. It was by using these sparks, and observing the alteration of the length of the spark when the spark-gap was placed in different positions, that Hertz demonstrated the existence of electric waves. He showed that they were reflected in the same way as light waves, that like these they could be brought to a focus, polarised, and could produce inter-ference effects from which their wave-length could be calculated. Younger physicists, using the very delicate methods for detecting electrical waves now available, will not realise the difficulty of these experiments, but older ones, who like myself began by using Hertz' method and had to observe whether tiny sparks, only a fraction of a milli-metre long, waxed or waned when the detector was moved from one position to another, will remember how arduous and harassing these experiments were and how long it took to make sure that the effects observed were not spuri-ous.[1] Rough as the method seemed, it was able in Hertz'

[1] A remarkable example of the difficulty of being sure that small changes in the intensity of light are real and not subjective is that of the N rays, which attracted a great deal of attention at the beginning of this century. They were first brought into notice by a distinguished French physicist, M. Blondlot, who came to the conclusion that the light from a Welsbach mantle, or Nernst filament, was accompanied by a species of radiation which, like Röntgen rays, could pass through substances opaque to ordinary light but, unlike them, could be refracted. He called them N rays. These rays did not produce light by themselves, but when they fell on a faintly luminous source of light, such as a feeble electric spark or a dim patch of phosphorescence, they altered the intensity of the light, generally, though not always, increasing it. A great many physicists tried to verify these effects, but for the most part failed to do so. It was remarkable that the power of seeing N rays seemed to be confined to

hands to prove the existence of electrical waves and to enable him to establish the theory which Maxwell had put forward. The names of Maxwell and Hertz will always be associated in the history of this subject. It is remarkable that of Hertz' many experiments, the one that did most to convince people of the existence of these waves was that which was supposed to be analogous to the production of stationary waves in optics or acoustics. This, however, turned out not to be the case. In the optical waves the distance between the nodes and the loops depends only upon the wave-length of the waves. It was found, however, that this distance in Hertz' experiment depended to a large extent upon the size of the instrument used to detect it. This was traced to the emission of waves by the detector itself when struck by the incident electrical waves.

Another name that should be mentioned in association with those of Maxwell and Hertz is that of Sir Oliver Lodge, who made many very beautiful and striking experiments illustrating the vibrations excited when a Leyden jar is discharged. The time of these oscillations depends on the length of wire connecting the inside and outside of the jar, and the jar can be tuned by altering this length. If a jar is connected with an induction coil,

Frenchmen ; as far as I am aware no English, German or American physicist succeeded in finding them, while in France they seemed to be universal. Some observers thought they had detected them coming from human beings, from plants and frogs, as well as from the luminous flames used in the earlier experiments. An incident which occurred at a demonstration of the rays seemed clear evidence that they were subjective (*Nature*, September 29, 1904). An audacious spectator, taking advantage of the opportunity afforded by the darkness of the room, managed to twist the aluminium prism which was supposed to direct the rays on to the place where they were to be observed, into quite a different position. The demonstrator continued to locate the rays in the place they were before the prism was moved. I believe it is now generally accepted that the effect is a subjective one, that the proverb " Seeing is believing " is true for them, though the interpretation is not the usual one.

and another jar in its neighbourhood is without any such connection, this jar can be tuned by altering the length of wire between the outside and inside, so that it sparks when the other coil is in action ; when it is thrown out of tune by altering the length of the wire, the sparks stop. He had also, almost simultaneously with Hertz, obtained evidence of electric waves ; his waves, however, travelled along wires, while Hertz observed the " wireless " waves travelling through free space. These are the waves which now play a large part in the lives of many millions of people in this country alone. As an example of how the practical applications of any scientific discovery may exceed the expectations even of those best qualified to judge, I may mention that when the Marconi Company was being formed Lord Kelvin told me that he had been asked to join the board of directors and that he had said that he would do so on two conditions. One was that Oliver Lodge should also be asked, and the second that the capital of the company should not exceed £100,000, as that, in his opinion, was the maximum amount that could profitably be employed in wireless. I imagine he thought that the most important use for it would be for communication between ships at sea and the land.

Argon

The discovery of Argon is one of the most romantic episodes in the history of physics, for it established the fact that there is in the atmosphere a gas which had entirely escaped notice, in spite of investigations by chemists on the composition of the atmosphere, extending over more than half a century. This is all the more remarkable because the amount of argon in the air is very large, being as much as 1·3 per cent by weight. In a cubic yard of air at atmospheric pressure there are about 13 grammes of argon, so that even a moderately-sized room will contain pounds of this gas ; several grammes of it are passing through our lungs each hour. A very interesting, impartial and detailed account is given in the *Life of Lord Rayleigh*, by his son. It reads almost like a detective story. It begins by Lord Rayleigh finding in 1892 that nitrogen prepared from the atmosphere was heavier, volume for volume, than that prepared from chemical compounds. This suggested that there was some intruder lurking in the atmosphere ; the problem was to detect him. The first thing to do was to make sure of the facts. Lord Rayleigh prepared the chemical nitrogen from several different kinds of chemicals, but they all gave the same result. Thus the intruder must be in the atmospheric nitrogen, and it must be heavier than the nitrogen. The question was, how did it get there and what was it ? Did it come in the preparation of atmospheric nitrogen ? To get nitrogen from the air, the oxygen and carbonic acid gas must be cleared out. These are processes which act, so to speak, as chemical charwomen and clear away all the litter from the nitrogen. Had they introduced the villain of the piece ? This did not seem possible, for several kinds of charwomen were tried and they all left the same kind of nitrogen. The thing to do next was to take away the real nitrogen and see if anything was left. Cavendish long ago had shown that this can be done by mixing the nitrogen with oxygen and sending electric sparks through the mixture. The nitrogen and oxygen combine and the compound they form can be removed by a spray of caustic potash. Rayleigh used this method, which, with the apparatus at his command, went on very slowly and was very laborious, and it required constant attendance, since the apparatus for producing the sparks kept stopping and needed constant adjustment. The experiment went on far into the night, with a telephone arranged near it to transmit the noise to him in an adjoining room where he was dozing in an armchair. When the noise stopped he awoke and went to start the instrument again. The diminution in the volume of the nitrogen went on for eight or nine days, then became very slow, and ultimately stopped. No amount of sparking produced any further diminution and the residual gas, when examined by the spectroscope, was proved to be neither hydrogen, oxygen nor nitrogen. Thus the intruder was at last trapped and shown to be a perfect stranger. To make assurance doubly sure, Rayleigh tried the same experiment using chemical nitrogen instead of the nitrogen from the air ; but this completely disappeared, leaving no residue. Thus the substance obtained in the first experiment could not have been produced by the sparking. Thus, at long last, was captured the arch-eluder who had escaped detection, though his haunts had been searched continually by chemists since chemistry began. He had taken the measure of Scotland Yard but had forgotten Sherlock Holmes, and when he appeared the game was up. What made the discovery especially remarkable was that it was made by the use of the balance, an instrument which is, and has always been, in every chemical laboratory, and in the use of which chemists are very expert. It was not discovered, though it might have been, by the spectroscope as so many other elements have been. When the properties of argon came to be examined, they were found to be very different from those of any other element ;

roughly speaking, it had no chemical properties, it formed no compound with any other element, it would have nothing to do with the most tempting brides that the chemists put before it ; as this is the trap on which chemists rely for catching a new element, it is no wonder that argon eluded them. The discovery of argon was much more important than the ordinary discovery of an element ; it was the discovery of a new type of element, one which has been of fundamental importance in the development of the theory of the structure of the atom. Sir William Ramsay, who had been trying to isolate argon by absorbing nitrogen by magnesium, and who had succeeded in doing so within a few days of Rayleigh's success with the sparking, afterwards discovered helium, neon, krypton, and xenon, four other elements of the same type as argon. The first announcement of the discovery of argon was made by Rayleigh and Ramsay at the meeting of the British Association at Oxford on August 13, 1894. It was an oral statement and Rayleigh was the speaker. This was never published, though some account of it appeared in various newspapers. The definitive account of the research was given in a paper read before the Royal Society on January 31, 1895. After this the existence of a new gas in the atmosphere was definitely admitted, though rather grudgingly in some quarters. It was quite natural and reasonable that chemists should be sceptical about the existence of a gas which was present in large quantities in the atmosphere but which had never been detected by any chemist, and whose properties seemed to be contrary to those of other gases. Thus argon, though its density is greater than nitrogen, is not so easily liquefied, while in general the heavier gaseous elements are more easily liquefied than the lighter. This fact seemed conclusive to

Dewar, Rayleigh's colleague at the Royal Institution, who was busily occupied in the liquefaction of gases, and made him at first a vehement opponent of the idea of the existence of the gas. In addition, there was never any love lost between Dewar and Ramsay. Criticism was to be expected, but what many resented was the contemptuous way in which some chemists spoke of the work, and dismissed the existence of argon as impossible because the instincts of a trained chemist warned him that it could not be true.

Röntgen Rays

The discovery of argon was formally announced at a meeting of the Royal Society in January 1895 ; before the year had ended another discovery of primary importance (that of the Röntgen rays) had been made, one which was destined to prove of first-rate importance for the study of the structure of atoms and matter. The first discovery of these was not, like that of argon, the result of long and laborious investigations; it came almost by chance. One day in the autumn of 1895 Professor Röntgen was experimenting in his laboratory at Würzburg to see if a Crookes tube, containing gas at a very low pressure, would give out invisible light if a current of electricity went through it. He had covered up the tube with black paper to shield off the visible light, and was amazed to find that when he sent the current through the tube a piece of cardboard covered with powdered fluorescent substance shone out with a glow bright enough to be easily visible. Something must have come out of the tube, got through the black paper and reached the fluorescent screen. Ultra-violet light of any known type could not have done this, as it is more easily absorbed than visible light. Shadows were cast on the fluorescing substance when a body was placed between the bulb and the screen. The shape of the shadow showed that the rays producing it travelled in straight lines from the bulb, and that they started from the place where the cathode rays in the tube struck against the anode, which in this tube was right opposite the cathode. It was soon found that the rays from the tube affected a photographic plate [1] so that a permanent record of the shadows cast by any object could be obtained. Dense objects cast a blacker shadow than light ones, so that in, e.g., the shadow of a hand, the bones stood out very distinctly against the flesh. Indeed, when the pressure in the tube was very low, the flesh could hardly be distinguished on the photograph. Similarly, the shadow of a purse showed the money in it, but not the purse itself. The importance of this property of the rays for surgery was at once realised by doctors ; indeed Röntgen's first paper on his rays was read before the Medical Society at Würzburg.

We organised a scheme at the Cavendish Laboratory by which photographs of patients brought by doctors were taken by my assistants, Mr Everett and Mr Hayles. The results were sometimes disconcerting to the doctors. They often showed that broken bones had not been set properly, and that the two ends were separated by a wide interval and connected only by a callus. One of the patients was a prominent member of the University who could express himself strongly. He had broken his arm and, as it did not heal, he insisted on having it Röntgen-

---

[1] One observer had noticed that his photographic plates got fogged when they were near a discharge passing through a gas at low pressure. All he did was to move the plates further away ; he saved his plates but lost the Röntgen rays.

rayed and brought his doctor with him. My assistant came back after a shorter time than usual. I asked him why. He said, "I came away because I thought the doctor would not like me to hear what Mr —— was saying to him after he saw the photograph".

A Röntgen-ray installation is now part of the equipment of every hospital. Few have done more to relieve human suffering than Röntgen and those who, by developing the application of the rays to surgery, have supplied the surgeon with his most powerful means of diagnosis.

The rays, as they well might, aroused much general interest. They seemed a long way towards the fulfilment of Sam Weller's need for " a pair of patent double million magnifyin' gas microscopes of hextra power—able to see through a flight of stairs and a deal door".

Never, perhaps, has a discovery of first-rate importance been so quickly and so abundantly verified as that of the Röntgen rays. The apparatus required was so simple (an induction coil, a Crookes tube and a photographic plate) that it was at hand in every physical laboratory, and there were few of these in which experiments on the rays were not started. Many interesting properties were discovered, often practically simultaneously in different laboratories. At the Cavendish Laboratory we soon found that when the rays pass through a gas they make it a conductor of electricity ; they produce positively and negatively electrified particles in the gas which move under the action of an electric force. I had been studying the conductivity of gases under electric forces for some time, but had been much hampered by the difficulty of finding any method of producing the conductivity which was efficient and reliable. The new rays removed this difficulty : they could be applied to gases at all temperatures and pressures, and could be standardised by observing how much they were diminished in intensity by passing through a sheet of tin-foil of definite thickness. Experiments which had been impossibly difficult before the discovery of the rays became easy.

We made at the Cavendish Laboratory many attempts to see if the rays which had passed through a thin plate of tourmaline would, as light would, show traces of polarisation, but never succeeded in finding any such effect. The polarisation of Röntgen rays was proved some years later by Professor Barkla. Many researches were made to try to distinguish between various views which had been put forward as to the nature of the rays. The views which received the most support were :

(1) That they were longitudinal vibrations in the ether. The argument in favour of this was that if they were vibrations in the ether, they must be longitudinal, otherwise they would show traces of polarisation. As soon as polarisation was discovered this argument lost all its force.

(2) That the rays were a form of ultra ultra-violet light, the wave-lengths being very much less than any ultra-violet yet discovered. It was not a valid objection to this view that, while ultra-violet light was both more absorbed and more refracted when passing through matter than visible light, the Röntgen rays behaved in quite the opposite way. The absorption and refraction of visible light by matter may depend upon resonance between the vibrations of the light and those of the matter, and this would disappear if the vibrations of the light were much more rapid than those of the matter. The view now generally held is that Röntgen rays may be regarded as ultra-violet light, stress being laid on the ultra.

(3) Another view, first put forward by Stokes, was

that Röntgen rays are thin pulses, differing from light as the sound of a flash of lightning differs from the roll of the thunder to which it gives rise. The absence of refraction was, in his view, due to the shortness of the time the forces in the thin pulse acted on the molecules of the refracting medium, that this was so short that only an infinitesimal amount of energy would go from the pulse into the refracting substance, too little to affect the velocity of its propagation. If, instead of a pulse, a long train of waves were passed through the substance, a state of equilibrium might be expected to be reached, in which a finite amount of energy went into the refracting substance and would give to it its refracting power. I worked at the problem from the other end and calculated what effects would, on Maxwell's theory, be produced if a moving electrified charge were suddenly stopped. This calculation showed that a pulse of electric and magnetic forces would be produced, that the energy in it would be inversely proportional to its thickness, which would be the distance light would travel in the time taken to stop the charged particle. The energy flowing through a closed surface surrounding the electron would be a/d times the energy possessed by the electron before it was stopped, where a is the radius of the electron and d the thickness of the pulse.

The atom on modern views is not an impenetrable sphere, but a positively electrified nucleus surrounded by electrons, the volume occupied by the electrons and the nucleus being an exceedingly minute fraction of the volume in which they are enclosed ; in fact, the atom is chiefly holes. Under these circumstances it is clear that it is a matter of chance how long a collision may last, and therefore how thick the pulse it produces may be. Thus the radiation given out by a substance bombarded by cathode rays is analogous to the radiation given out by an incandescent body, and consists of a mixture of light of different wave-lengths, the proportions of the different waves being determined by statistical considerations. One thing is clear, that the energy in a pulse cannot be greater than the energy in the cathode ray which produces it, and since the energy in the pulse is inversely proportional to its wave-length, there will be a lower limit to the wave-length of the Röntgen rays given out by bombardment by cathode rays ; at this limit the energy in the pulse is equal to that possessed by the fastest cathode ray striking the target. A few years later an old student of mine, Professor Barkla, now Professor of Physics at the University of Edinburgh, showed that the resemblance between Röntgen rays and light was even more complete than had been realised, and he showed that Röntgen rays could be polarised and, what was even more important, that each chemical element gave out a characteristic Röntgen ray spectrum when bombarded by cathode rays whose energy exceeded a certain limit. This was in addition to the continuous spectrum corresponding to thermal radiation. If, for example, we bombard silver with cathode rays, beginning with those of small energy, at first the only radiation is the continuous one, but if we increase the energy of the cathode rays we reach a stage when this is supplemented by a Röntgen radiation (A) of definite wave-length ; continuing to increase the energy of the cathode ray, we get for a time the continuous radiation and (A), but when the increase has reached a definite value we get a new type of radiation (B), having a smaller wave-length than (A), and by increasing the energy we may get other types of radiation. The radiations (A), (B) were called by Barkla the characteristic radiation of the substance bombarded.

They correspond to the lines in the visible spectrum of the various elements. This was a discovery of fundamental importance which was most deservedly rewarded by the award of a Nobel Prize.

Diffraction of Röntgen Rays

In the diffraction gratings used for spectroscopy the rulings are not usually separated by less than 10^{-4} cm., while the wave-length of the characteristic Röntgen radiation of silver, for example, is at only about $\frac{1}{20,000}$ of this. It is evident that it would be as likely to get diffraction of Röntgen rays from such a grating as to get diffraction of light from a grating with the rulings 1 cm. apart. Laue, however, realised that, since in a crystal the atoms are arranged periodically at intervals equal to the distance between two molecules, a crystal ought to act as a diffraction grating for wave-lengths comparable with this distance. He directed a beam of Röntgen rays on a crystal, and obtained well-marked diffraction effects, from which he was able to calculate the wave-length of the Röntgen rays. These proved to be exceedingly small compared with the wave-length of visible light, thus the wave-length of the hardest characteristic radiation from silver is only about one ten-thousandth part of the wave-length of the D line of sodium. Moseley, a pupil working in Lord Rutherford's laboratory in Manchester, obtained the very important result that the square of the frequency of the hardest radiation given out by an element was a linear function of the atomic number of the element. From this law we can find the atomic number of any element whose characteristic radiation has been determined, and thus find if any elements remain to be discovered. This was one of the most brilliant discoveries ever made by so young a man, and science suffered a grievous loss when he fell in the war a few months after publishing his discovery.

Instead of using a crystal to determine the wave-length of a characteristic radiation, we may use the diffraction pattern given by Röntgen rays of known wave-length, to determine the arrangement of the different atoms in a molecule of a compound which exists in a crystalline form. This method, in the hands of Sir W. H. Bragg and his son Professor W. L. Bragg, has developed into a method of great power and importance. It is very satisfactory to find that in many cases it leads to the same results as the chemists had arrived at by purely chemical considerations.

The Röntgen rays naturally were the subject of many public lectures and of discussions organised by scientific societies. I was appointed Rede Lecturer at the University of Cambridge in 1896 and took them as the subject of a lecture to a very large audience. They naturally played the leading part in the proceedings of Section A at the meeting of the British Association in Liverpool in 1896, the first meeting after their discovery. I happened to be President of the Section and there was a large gathering of physicists, including Lord Kelvin, Sir G. G. Stokes, Professor Oliver Lodge, Professor G. F. Fitzgerald, and, as guests of the Association, Professor Lenard of Heidelberg and Drs. Elster and Geitel of Wolfenbüttel.

The discussion on Röntgen rays began with a paper by Professor Lenard, who was the first to detect rays outside a Crookes tube. These, however, were not Röntgen rays but Cathode rays which had passed through a window of very thin gold leaf in the tube. Sir George Stokes gave a beautifully clear and animated exposition of his theory of Röntgen ray.

Among the guests of the Association were Drs. Elster and Geitel, who furnish the most remarkable instance on record of a lifelong partnership in research in physics extending over forty years. They were boys together at the same school ; for thirty-nine years they were both teachers in the secondary school at Wolfenbüttel ; they lived the greater part of their lives in the same house, which had a well-equipped physical laboratory where their research was done, and though they were offered a dual university chair they preferred to stay at Wolfenbüttel. Their work covered a very wide range : conduction of electricity through flames, photo-electric effects, problems in phosphorescence and in meteorology. It was all of very high quality and forms a very valuable and substantial contribution to physics. Personally they were the most delightful companions, with charming old-world manners. They were both keen gardeners, and in their greenhouse they not only grew flowers, but also bred highly coloured butterflies to settle upon them.

It was at the Liverpool meeting of the British Association that Sir William Preece, the electrical engineer to the Post Office, announced that a young Italian inventor named Marconi had come over with a new method of signalling through space, and that the Post Office were giving him facilities for testing it. Marconi's first patent was taken out in 1896. The Liverpool meeting was also memorable as being the first meeting attended by Ernest Rutherford, who had come to Cambridge a short time before as a research student.

As Röntgen rays are light of very small wave-length

the nature of light ought to be the same as that of these rays. Röntgen rays ionise a gas through which they pass, and if the wave front of a beam of these rays were continuous no molecule in the path of the beam could escape from being struck by the rays and all would be equally affected by it. We should expect either that no molecules would be dissociated or that a large proportion would be. This, however, is not the case. In the strongest beam of Röntgen we can produce, only an infinitesimal fraction of the molecules struck by the beam are ionised. I pointed out in the Gilliman Lectures that I gave at Yale University in 1903 (*Electricity and Matter*, Scribner) that this indicated that the front of the beam could not be continuous, but must be more like a series of bright spots on a dark background, *i.e.* that the energy must be concentrated in separate bundles. This is a small part of what was afterwards known as the Quantum Theory of Light, the other part of that theory being Planck's Law that the energy in each bundle is equal to $h\nu$.

The quantum of light must, like a corpuscle on the Corpuscular Theory of Light, travel through space without change, and yet it must, like the train of waves of the Undulatory Theory, produce interference phenomena. I suggested in a letter published in *Nature*, February 8, 1936, that the quantum *is* a train of waves, but that the waves are of a somewhat different character from those in the ordinary theory. The waves I considered were waves where the lines of electric force were circles, all these circles had their centres on a straight line and their planes at right angles to it. I showed that a train of waves of this kind would travel out in the direction of its axis without suffering any change, thus satisfying the first condition for the quantum, while, since the quantum is a train of waves, it would produce interference effects.

Radio-activity

" It never rains but it pours." A few months after the discovery of Röntgen rays, Becquerel found that salts of uranium, either when crystalline or in solution, affected after a long exposure a photographic plate protected by a covering opaque to ordinary light, and that they emitted radiations which, like " Röntgen rays when they passed through a gas, made it a conductor of electricity ". Rutherford, working at first in the Cavendish Laboratory, investigated the radiation from uranium very thoroughly. He found that the radiation was of two types, one type, which he called the alpha type, being absorbed after passing through a few millimetres of air, while the other, the beta type, could get through more than twenty times this distance. He used the electrical method of investigating the intensity, measuring this by the ionisation it produced in a gas. Uranium is the element of greatest atomic weight. Shortly after Becquerel's discovery Schmidt found that the element next in atomic weight, thorium, possessed similar properties.

But most important of all was the discovery by Monsieur and Madame Curie of a new element, radium, which possessed radio-active powers enormously greater than those of uranium and thorium. Madame Curie had made a systematic examination of a great number of chemical elements and their compounds, and also of minerals, to see if they could find other elements exhibiting radio-active powers. The investigation of the elements and their compounds did not lead to anything new. The investigation of the minerals yielded, however, surprising results : it was found that several minerals containing uranium were more active than the same bulk of pure uranium. This suggested that there was something more radio-active than uranium in these minerals. They proved this in yet another way. They prepared from pure substances a body which had the same chemical composition as the mineral chalcolite, which is very radio-active, and found that this had only one-fifth the activity of the native substance.

M. and Mme Curie set to work to isolate the substance or substances responsible for the very great radio-activity of the pitchblendes, the ores from which the greater part of the uranium used in commerce is extracted. After great labour they succeeded, in 1898, in extracting from about two tons of pitchblende residues one-tenth of a gram of an intensely radio-active substance which they called radium. The amount of radium in the ore is so small, about one-thirtieth of a pennyweight per ton, that its separation would have been impossible if the radiation from the substance itself had not supplied a means for its detection far more sensitive than any known chemical or spectroscopic test. Even so, the separation required much time and work, and as M. and Mme Curie had very small means and had to rely upon what they earned by teaching, it was a hard struggle for them to find the time and money required for this investigation. Having got the radium, they lost no time before investigating its properties, M. Curie taking the physical and Madame the chemical, and for many months hardly a number of the *Comptes Rendus* appeared without the announcement of some new and striking property of radium. One made by MM. Curie and Debord was that the temperature of the radium salt was always higher than the surrounding atmosphere, show-ing that the radium was itself a source of heat. The dis-coverers of radium, as well they might, soon received recognition from learned socieites. In 1903, M. and Mme Curie received the Rumford Medal from the Royal Society and shared with M. Becquerel the Nobel Prize. In April 1906, M. Curie met with a tragic and fatal accident. He was knocked down in Paris by a cab, and a lorry passing at the time ran over his head. He was then forty-seven years of age. In addition to his work on radium, he had made investigations which are classical on the variations of the magnetic properties of bodies with temperature. He discovered that the coefficient of magnetisation of magnetic substances is inversely proportional to the absolute tem-perature, a result which is always called Curie's Law. He made also important investigations on piezo-electricity, and it is noteworthy that the electrical measurements which made the detection of radium possible were all made with an electrometer whose action depended on this property. He came to us at Cambridge once when he was visiting England : he was the most modest of men, ascribing every-thing to his wife, and had a most attractive simplicity of manner.

Let us, however, return to Rutherford's experiments in the Cavendish Laboratory on the radio-activity of uranium and thorium. Those on uranium gave comparatively little trouble; the results got on one day could be repeated on the next. The behaviour of thorium, on the other hand, was most perplexing and capricious ; changes in the surroundings which seemed quite trivial, such as opening a door in the workroom, produced a very large diminution in the radiation, while, on the other hand, very large changes in the physical conditions produced no appreciable effect upon it. The radio-activity seemed to act like a contagious disease and infect solid bodies placed near the thorium. These, however, recovered in time if the thorium were taken away. These vagaries turned out, however, to be an illustration of the principle that difficulties in the experiments may be the seed of great discoveries, for Rutherford, when he went as Professor of Physics to Montreal, resumed these experiments and, in his attempts to unravel their intricacies, was led to a discovery of fundamental importance which was the origin of modern views about the processes going on in radio-active substances. This discovery was that thorium gives off something which he called an emanation, which is itself radio-active, and which is in the gaseous state, and can thus be wafted about by currents of air, and may settle on solids and make them behave as if they were radio-active themselves. The radio-activity of the emanation is not permanent, but only lasts for a few hours, at the end of which the emanation has passed into another non-radio-active substance. The emanation behaves like the inert gases helium and argon : it does not enter into any chemical combinations, and the length of its life is not affected by any influence external to its molecules. High temperatures, or even bombardment by the rays given out by strongly radio-active substances such as radium, do not seem to do it any harm, nor does it last longer when radiation from outside is warded off by surrounding it with thick layers of lead. Thus the thorium emanation is an element which has a transitory life ; it is liable to be seized with a convulsion and give out an α ray, *i.e.* an atom of helium with a double positive charge, and when it has done this it becomes another element with smaller atomic weight and different properties. This new element may itself after a time change into another. The average lives of these elements vary very greatly ; for some it is only a fraction of a second, for others it may be many years.

On the view that an atom consists of electrons revolving round a nucleus, each electron makes in one second more than a million million million revolutions, so that, reckoned on a time scale fixed by the time of one of these revolutions, a second would be an eternity. To explain phenomena occurring at so long an interval, we must find some time interval, occurring in an isolated atom, which may be comparable with a second or even with many years. We cannot do so if there is only one electron in the atom —but suppose there are two. Sometimes these may be in a straight line with the nucleus, but if their periods of rotation are very nearly equal, each will have to make a very large number of revolutions before they are again in a straight line. The interval between two such configurations gives a time associated with this atom much longer than that associated with an atom containing only one electron. In an atom containing many electrons there are possibilities for configurations containing several electrons ; the more electrons there are in a configuration the longer will be the interval between its recurrence, and the increase may be very rapid. Now it is possible that the reactions between the electrons and the nucleus may be such that for a particular configuration of the electrons—for example, if all of them were pulling in the same way—the nucleus might be distorted from its original stable position to an unstable one, and fall from this to another stable position, exploding, in fact, and emitting as it does so a large amount of energy, which is carried away by one of the constituents of the nucleus, the α-particle. The interval between two explosions, or rather the average time before an explosion takes place, may from the preceding considerations be of quite a different order from the time of revolution of an atom.

Wilson Tracks

The method of investigating the nature of various kinds of rays—a rays, β rays, γ rays, cosmic rays—by " Wilson tracks ", which is now of paramount importance in the study of atomic structure, was invented and developed by Professor C. T. R. Wilson, F.R.S., at first in the Cavendish Laboratory and after the war at the Solar Physics Observatory, Cambridge, in a research which lasted for more than twenty years. He began shortly after taking his degree in 1892 by experimenting on fogs, not, it might seem, a very obvious method of approaching transcendental physics. He studied the conditions under which fogs are formed. It was known that fogs are not formed under normal conditions unless dust is present. This is due to the effect of the surface tension on the evaporation of drops of water ; this promotes evaporation, and will make a drop evaporate when surrounded by air saturated with water vapour. The effect increases as the size of the drop diminishes. It is evident that this property makes the growth of drops from smaller ones of microscopic dimensions impossible, for these small drops would evaporate and get smaller, and the smaller they get the faster will they evaporate. They are in the position of a man whose expenditure increases as his capital decreases, a state of things which will not last long. If particles of dust are present, the water can condense on their surface, and the conditions at its surface are the same as for a drop of water the size of the particle of dust ; for particles as large as this a small supersaturation will be sufficient to prevent evaporation, and when more water is deposited and the drop gets larger, the evaporation will get less and less; the dust enables the drop to short circuit the impassable early stages.

If the drop is electrified, the effect of the electrification is the opposite to that of surface tension. An electrified drop will not evaporate even if the air around it is not saturated, and the smaller the drop the smaller is its tendency to evaporate, and the greater the deposition of moisture from the air outside. For drops below a certain size the electrical effect will be greater than that of surface tension, and a drop, if formed, will increase until it reaches the critical size. Thus drops should be formed even in the absence of dust if they are electrified.

In Wilson's experiment, the air in which the fog is produced was contained in a vessel one of whose boundaries was a movable piston which could move vertically up or down. When the pressure below it was suddenly reduced, the piston was very rapidly pushed downwards so that the volume of the air in the vessel above the piston was suddenly increased. This adiabatic expansion cooled the air in the tube, and since the air in the tube was, by being in contact with water, saturated before the expansion, it was supersaturated after it. Before the dust was removed from the air a small expansion was sufficient to produce a fog; as the fog settled it carried some dust down with it, and by repeated expansions the dust could be removed. When this was done, no fog was produced by expansion which produced six- or seven-fold supersaturations, but when the gas in the tube was ionised by exposure to Röntgen rays or to the radiation from radium, a dense cloud was produced by a four-fold supersaturation. This appeared to be uniformly diffused through the tube.

Wilson then made a step which was all-important for the purposes for which the cloud method has proved so valuable. Instead of looking at the tube under continuous illumination, he took instantaneous photographs of the tube at a very small interval after the expansion, so small that the ions had not time to move more than a very small distance from the place where they were produced. If the ions were produced by a particle like an a or β particle passing through the tube, they would be produced along the track of the particle, and the drops of water deposited on these ions would be all on this track, and a photograph of these drops would represent the path of the particles.

In order to get a stereoscopic effect he took two photographs simultaneously, using two cameras placed in different positions. Specimens of these photographs are given in Figs. 1, 2 and 3, Plate II. In Fig. 1 the ionisation was produced by an a particle. Here the ionisation was so great that the path appears continuous : on one of these there is a sharp bend showing the a particle had been deflected by collision with a molecule of the gas through which it was passing. Fig. 2 shows the path of a fast β particle through the gas : here the ionisation is not nearly so great as for the a particle and the separate drops can be distinguished ; the drops are farther apart with fast β particles than with slow ones, confirming the result indicated by theory, that ionisation by moving electrified particles diminishes with the velocity of the particle after a certain velocity is exceeded. If a magnetic force is applied to the particles they will no longer move in a straight line but will describe a curve, and the photographs would show curved lines instead of straight ones : if the particles were negatively charged like electrons they would be bent in one way ; if they were positively charged, in

PLATE II

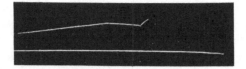

FIG. 1

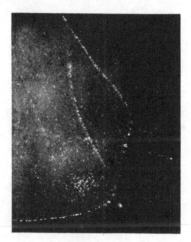

FIG. 2

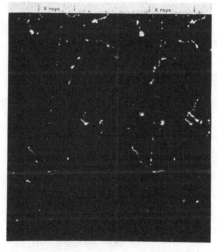

FIG. 3

the opposite. It was by observing that in some of his experiments there were particles bent in opposite directions to about the same extent, that Anderson discovered positively charged particles having much the same mass as the electron. His results were confirmed and extended by the work of Blackett and Occhialini. Fig. 3 shows the state of things when the gas is ionised by Röntgen rays. Here it will be seen that instead of a single line we have a very large number of lines starting from places in the gas : this shows that the greater part of the ionisation by Röntgen rays is produced by charged particles which are ejected from the molecules of the gas when these are struck by Röntgen rays.

This work of C. T. R. Wilson, proceeding without haste and without rest since 1895, has rarely been equalled as an example of ingenuity, insight, skill in manipulation, unfailing patience and dogged determination. Those who were not working at the Cavendish Laboratory during its progress can hardly realise the amount of work it entailed. For many years he did all the glass-blowing himself, and only those who have tried it know how exasperating glass-blowing can be, and how often when the apparatus is all but finished it breaks and the work has to be begun again. This never seemed to disconcert Wilson ; he would take up a fresh piece of glass, perhaps say " Dear, dear ", but never anything stronger, and begin again. Old research students when revisiting the Laboratory would say that many things had altered since they went away, but the thing that most vividly brought back old reminiscences was to see C. T. R. glass-blowing. The beautiful photographs that he published required years of unremitting work before they were brought to the standard he obtained. The method has been quickened up and made automatic by other workers, but though they can turn out more photographs in a given time, the photographs themselves are no better than those got by C. T. R. more than twenty years ago. It is to him that we owe the creation and development of a method which has been of inestimable value to the progress of science.

Lord Kelvin

Any notice about the progress of physics in the latter part of last century would be like the play of *Hamlet* without the Prince of Denmark if it did not deal with the part played in it by Lord Kelvin, who for more than forty years before his death in 1907 had been the most potent influence in British physics. Of the 661 papers he published, those on Thermodynamics which have already been alluded to would, I think, in the opinion of most people, be regarded as the most important. This, however, was not his own opinion, for he told me that before the discovery of radium had made some of his assumptions untenable, he regarded his work on the Age of the Earth as the most important of all. In this he calculated the loss of heat by the earth due to the radiation from its surface of the heat coming up by conduction from the warmer parts below. He went into the question as to whether the earth could have any internal source of energy due to the combustion of the substances of which it was composed, and found that if these were known elements it would only account for a small fraction of the heat lost by radiation. He concluded that " it is highly improbable in the present state of science that any effect of this kind could compensate for the loss of heat by radiation ", and in that case the time between now and the solidification of the earth's crust could not be very much greater than 100 million years. This view was attacked vigorously by both geologists and biologists, by the first because it was inadequate for the deposition of the sedimentary rocks, and by the second because it was not sufficient for the development from some form of primordial life of the highly developed forms of life now existing. On his assumptions his conclusions were legitimate, but the progress of science makes things which seem highly improbable in one generation become firmly established facts in the next. The suggestion that bodies could exist with the power of emitting such enormous amounts of energy as do radium and other radio-active substances would have seemed, to say the least, "highly improbable" at the time (1862) when Kelvin's paper was written, but within fifty years the Curies had discovered radium and Rutherford had shown it would materially increase the life of the habitable earth. It is interesting to think that but for radio-activity the world would not be habitable. There are other sources of energy which might, like radium, prolong the life of the world, e.g. the transformation of protons into positrons would liberate large amounts. This, however, would not be spontaneous. Since the proton is a stable system it would require the expenditure of a considerable amount of energy, although but a small fraction of that finally liberated, to start it.

Thomson's services to science received early recognition. He was only twenty-two when he was elected to the Professorship of Natural Philosophy in Glasgow. He was elected a Fellow of the Royal Society in 1851 on the same day as Stokes and Huxley (this was before the publication of his important researches on Thermodynamics).

He spent a great deal of time between 1856 and 1866 on work associated with the laying of the Atlantic cable between England and America. His connection with this arose from his becoming in December 1856 a director of the Atlantic Telegraph Company, formed to lay a cable between this Company and the United States. He threw himself into this work with characteristic energy and enthusiasm; he solved the problem of finding the law which determines the rapidity with which signals can be sent through a cable; he showed that though the current at the transmitting end is only kept on for a short time, the effect at the receiving end will last far longer. Unless the interval between signals from the transmitting is greater than the time for their effects at the other end to die away, one signal will get confused with the one before it, and the messages will get so blurred that they are not intelligible. To detect the signals he invented the mirror galvanometer, which was far more sensitive than any which had been used before. The first attempt to lay the cable was made in June 1858. Thomson was in one of the cable-laying ships, but he was not responsible for the apparatus and methods used as these were under the control of Mr Whitehouse, the electrician appointed by the Company. The cable was laid in spite of many difficulties due to heavy storms, and messages were sent through it; one from Queen Victoria of ninety-nine words took $16\frac{1}{2}$ hours. The messages coming through the cable got feebler and feebler and stopped altogether after about three weeks. A second attempt was made several years later when fresh capital had been raised, and in this case Thomson was the electrician in charge. A new cable was made with the copper core larger than in the earlier one and its insulation more carefully tested. At the first attempt to lay it, it parted in mid-ocean, where the depth was 2000 fathoms. It was subsequently recovered, but meanwhile another cable had been made, and was successfully laid by September 1866. This, as it well might, created immense enthusiasm in the country, and Thomson's fame spread throughout the length and breadth of the land. He was created a knight. He had done a work which could only be done by one who was at once a great physicist and a skilled electrical engineer, and who had great determination, driving power and enthusiasm. Thomson was not only enthusiastic himself, his enthusiasm was contagious. When he visited the Cavendish Laboratory he would go round and see what researches were being done by the students. If there was an experiment in progress he would take keen interest in it, get the student to repeat it, and make suggestions for other experiments. All this was a great stimulus to the students and made for the success of the laboratory. His triumph with the Atlantic cable led naturally to his service being retained as Consulting Engineer by many cable companies. He had also, in partnership with Varley, a practice as Consulting Engineer, and he also appeared from time to time in the Law Courts as an expert witness. These all made additional demands on his time and caused many interruptions in his scientific work. He told me once that he relied on his journeys between London and Glasgow for thinking out any physical problem on which he was working.

Few men have obtained great distinction over such a wide range of subjects. He was a great physicist and had left his mark on almost every branch of physics known in his time. He was a great mathematician, and used his physics to guide his mathematics and his mathematics to give precision to his physics. He was also a great electrical engineer. He had been President not only of the Royal Society but also of the Mathematical and Physical Society, of the Institute of Electrical Engineers and of Naval Architects. Besides writing hundreds of scientific papers he had taken out more than sixty patents.

His mind was extraordinarily fertile in ideas. This was very evident at the meetings of Section A of the British Association, which he sedulously attended, accompanied by Lady Kelvin with a packet of sandwiches in case the meeting should be too prolonged. He would often get up after some paper had been read and say something quite new and always interesting, whether one agreed with it or not. If Professor G. F. Fitzgerald, who was also full of ideas, and Sir Oliver Lodge, who was unrivalled in clear exposition, were also present, one saw the British Association at its best. Even in a lecture, if a new idea occurred to him, he would start off on a new tack. This made them very discursive and often very lengthy. It was, or used to be, the tradition that the Friday Evening Discourses at the Royal Institution should not last for more than an hour, and I have known Sir James Dewar, when he was Director, walk up to a lecturer who had exceeded the hour by ten minutes and say to him, " You *must* stop ". Lord Kelvin, however, never paid any attention to this rule. He has been known to have lectured for the hour before reaching the subject of the lecture. It was only very rarely that he prepared either a lecture or a speech. There was, to the few who were already interested in the subject he was talking about, generally both charm and interest in these diversions. On the other hand, when he was writing an important paper he took almost meticulous care in choosing the word which would most clearly express his meaning. He was not fond of reading scientific literature and relied to a large extent on conversations with friends for information about new discoveries. In fact he was a remarkable exception to that very important physical principle that good radiators are good absorbers.

His first paper appeared in 1841, and from then until his death in 1907 there was no year without a paper, and few without many. They cover all branches of physics and record some of the greatest discoveries made in his generation. He himself, however, was not satisfied, for at his jubilee in 1896 he said, "One word characterizes the most strenuous of the efforts for the advancement of science that I have made perseveringly during fifty-five years, and that word is FAILURE. I know no more of electric and magnetic forces or of the relation between ether, electricity and ponderable matter, or of chemical affinity than I knew and tried to teach to my students of natural philosophy fifty years ago in my first session as Professor." Science never had a more enthusiastic, stimulating or indefatigable leader.

Niels Bohr

At the end of 1913 Niels Bohr published the first of a series of researches on spectra, which it is not too much to say have in some departments of spectroscopy changed chaos into order, and which were, I think, the most valuable contribution which the quantum theory has ever made to physical science ; most of them, however, belong to a more recent period than that covered by my survey.

Relativity

The germ of the theory of relativity was the experiment made by Michelson and Morley in 1887 to see if they could detect any optical effects due to the motion of the earth round the sun. There was reason to believe that such an effect might exist, for Maxwell had pointed out that if a ray of light starting from a point A travelled in the direction in which the laboratory was moving, and was reflected back by a mirror M at right angles to its path, the time lag between leaving and returning to A would be affected by the motion of the system. For suppose l is the distance between A and the mirror M, c the velocity of light, u the velocity of the mirror; then when the light is approaching the mirror the velocity of light relative to the moving system is $c - u$, and the time taken to reach the mirror is $l/(c - u)$; on the return journey the relative velocity is $c+u$, and the time taken to get back to A is $l/(c+u)$; the time lag is the sum of these terms, *i.e.* $2lc/(c^2 - u^2)$. Consider now a beam travelling along AN nearly at right angles to the velocity u, and then reflected from a horizontal mirror at N. The direction of AN is adjusted so that the reflected beam and the moving mirror arrive simultaneously at A' ; the time lag between leaving A and returning to A' is $2AN/c$ and also AA'/u ; it follows from this that the time lag is $2ON/\sqrt{c^2 - u^2}$, O being midway between A and A' ; ON is the vertical distance of the mirror N from the place from which the beam starts. Writing L for ON we see that the difference in the time lags when both beams return to A' is equal to

$$\frac{2}{c\sqrt{1 - u^2/c^2}}\left(L - \frac{l}{\sqrt{1 - u^2/c^2}}\right) \qquad . \qquad . \quad (1)$$

In the experiments A was a semi-transparent mirror, which reflected some light and allowed some to pass through it. The horizontal beam was transmitted through the mirror A when it started, and reflected from it when it came back, the vertical beam was reflected when going out and transmitted when it came back : the two beams in this way become parallel and there will be a change in the interference bands they produce unless the expression (1) vanishes. In calculating the effect to be expected, u was taken to be the velocity of the earth due to its revolution round the sun, the velocity of the sun itself is neglected as it is supposed to be small in comparison. The apparatus would have detected a difference of lag less than one-twentieth of that represented by (1), but no effect at all was detected. This could not have arisen from mistakes in the measurements of l and L which happened to cancel out the real effect, for the instrument could be rotated through a right angle when L and l would be interchanged, and errors in their measurements if they masked the effect in one position would increase it in the other. G. F. Fitzgerald and Lorentz both suggested in 1892 that the absence of any effect could be explained if the apparatus contracted in the direction in which it was moving in the ratio of 1 to $\sqrt{1 - u^2/c^2}$; this as will be seen by (1) makes the effect vanish. This contraction could not be detected by measurements made in the laboratory, for the scales used to measure these lengths would contract in the same proportion, and a length which was equal to the length of

a metre scale if the system were at rest would still be equal
to it if the system were in motion. Since strains very
readily produce double refraction in bodies, Lord Rayleigh
in 1902 tried if he could detect double refraction due
to the earth's motion, but was unable to do so ; an experiment on the same lines but on a large scale was made
by Brace in 1904, but again the results were negative.
This contraction is the same as that indicated by Maxwell's
theory (see page 367). In 1905 Einstein published his
theory of relativity. The object of the theory was to
establish a relation between a phenomenon in a system
at rest with one in the same system when it is moving in
a straight line with a uniform velocity u. Suppose that
some phenomenon occurs in the system at rest expressed
by a relation between (x, y, z) the co-ordinates of a point
referred to fixed axes, and the time t. Then on Einstein's
theory this will involve a phenomenon in the moving
system which can be expressed by substituting for x, y, z
co-ordinates x', y', z', and for t a time t', connected with
x, y, z and t by the relations,

$$x'=\frac{x-ut}{\beta}, \quad y'=y, \quad z'=z, \quad t'=t-\frac{ux}{c^2}$$

where β is written for $\sqrt{1-\frac{u^2}{c^2}}$.

Solving these equations we see that they are equivalent to

$$x=\frac{x'+ut'}{\beta}, \quad y=y', \quad z=z', \quad t=\frac{t'+\frac{ux}{c^2}}{\beta},$$

so that the equations to transform from the fixed system
to the moving one are of the same form as those to trans-

form from the moving to the fixed, but the sign of u must be changed. Let us take as an illustration the effect of motion on the length of a rod. Let the length of the rod in the moving system be l, x'_1 the co-ordinate of one end, $x'_1 + l$ the co-ordinate of the other, then if x_1, x_2 are the co-ordinates in a system at rest, we have

$$x_1 = \frac{x'_1 + ut'}{\beta}, \quad x_2 = \frac{x'_1 + l + ut'}{\beta},$$

$$x_1 - x_2 = \frac{l}{\beta}, \quad \text{or} \quad l = \beta(x_1 - x_2),$$

since β is less than one, $x_1 - x_2$ is greater than lx, in accordance with the Fitzgerald and Lorentz contraction.

Einstein showed that it followed from the principle of relativity that the mass of a moving body varies as $1/\sqrt{1 - u^2/c^2}$, and that the energy is equal to mc^2 where m is the mass when its velocity is u.

The relation $t' = \{t - ux/c^2\}/\beta$ leads to some apparently very paradoxical results. Suppose that, in a system at rest, one event occurs at the time t_1 at the place x_1 and another at the time t_2 at the place x_2, and that t'_1 and t'_2 are the times at which they are observed in the moving system, we have

$$t'_1 - t'_2 = \{t_1 - t_2 - u(x_1 - x_2)/c^2\}/\beta.$$

Hence when

$$t'_1 - t'_2 = 0, \quad t_1 - t_2 = \frac{u}{c^2}(x_1 - x_2),$$

so that t_1 is not equal to t_2, hence the events which are simultaneous in the moving system are not so in the fixed one. We see too that $t_1 - t_2$ and $t'_1 - t'_2$ may have opposite signs, *i.e.* the event which was first in the fixed system may be last in the moving one. Thus, for example, if a woman after the birth of one child moved into a different neighbourhood before the birth of a second, she might appear to have had twins to an observer in a moving system, while to an observer in a system moving more rapidly the second child would appear to have been born before the first. One wonders what the mother would have thought of relativity. The paradox, however, ceases when we remember that, as an observer of an event at a distance does not observe it at the time it occurs, but only after the interval required by light to travel to him from the scene of the event, his knowledge of distant events must always be out of date. Thus if he is observing two events at two different places, one nearer to him than the other, then though the events occurred simultaneously at the two places, he would perceive that at the place nearer to him before he did that at the other. When the observer is moving, his distances from the two places change, and if he is moving so as to approach the place that was at first further away and recede from the other, when he came to the place where the distances were equal, then simultaneous events would take the same time to reach him and he would regard them as simultaneous, and when he went still further, the one that was at first perceived after the other would now precede it.

The results arising from the principles of relativity were so surprising, and its exposition by Einstein so masterly, that it excited the interest and admiration of mathematicians and also of the general public to an extent which had never even been approached by any other mathematical question. Lectures on it attracted large audiences, books about it became " best sellers ", and one was continually being asked by one's neighbour at a dinner party to explain in simple words what it was about. Once at the Athenaeum Club, Lord Sanderson, who was for long permanent Secretary at the Foreign Office, came to me and asked if I could help him. He said, " Lord Haldane has been to the Archbishop (Randall Davidson) and told him that relativity was going to have a great effect upon theology, and that it was his duty as Head of the English Church to make himself acquainted with it " ; he went on, " The Archbishop, who is the most conscientious of men, has procured several books on the subject and has been trying to read them, and they have driven him to what it is not too much to say is a state of *intellectual desperation*. I have read several of these myself and have drawn up a memorandum which I thought might be of service to him. I should be glad if you would read it and let me know what you think about it, before I send it to him." I said I did not think relativity had anything to do with religion but I would read his memorandum. I see from the biography of the Archbishop by the Bishop of Chichester that just about this time the Archbishop met Einstein at Lord Haldane's house, and asked him what effect he thought relativity would have on religion. Einstein said, " None. Relativity is a purely scientific matter, and has nothing to do with religion."

The theory of relativity deals with physical phenomena. If we take the view that the structure of matter is electric, these ought to follow from Maxwell's equations without introducing relativity. We have referred to examples in which various effects have been explained in this way, *e.g.* the contraction of a moving body and the variation of mass with velocity, before relativity was introduced. On this view it is reasonable to regard Maxwell's equations as the fundamental principle rather than that of relativity, and also to regard the ether as the seat of the mass momentum and energy of matter, *i.e.* of protons and electrons : lines of force being the bonds which bind ether to matter. In Einstein's theory there is no mention of an ether, but a great deal about space : now space if it is to be of any use in physics must have much the same properties as we ascribe to the ether ; for example, as Descartes pointed out long ago, space cannot be a void. Unless there was something in space there would be nothing to fix the position of a point, position would have no meaning and space could have no geometrical properties ; the same would be true even if it were filled with a perfectly uniform substance, for again there would be nothing to differentiate one point from another. Space must therefore have a structure. Again, there must be in space something which changes. If there were not there would be nothing to distinguish one instant from another, nothing to supply a " clock ". Nothing can travel through space with a velocity greater than light; this is the speed limit for traffic through it, but space could not enforce this unless it had some time to measure the velocity, and this requires something by which we could measure times. Again, the mass of a body increases as the velocity increases ; if the mass does not come from space it must be created. It would seem that space must possess mass and structure both in time and space : in fact it must possess the qualities postulated for the ether.

" Naturam expellas furca, tamen usque recurret."

It has been urged against the existence of the ether that so many different kinds of ether have been from time to time proposed. The same may be said about " space " : we have Einstein's space, de Sitter's space, expanding universes, contracting universes, vibrating universes, mysterious universes. In fact the pure mathematician may create universes just by writing down an equation, and indeed if he is an individualist he can have a universe of his own.

The kind of relativity I have considered, "special relativity" as it is now called, was Einstein's first theory and deals with electric and magnetic problems which can also be solved by Maxwell's equations. Einstein has given a second theory, known as "general relativity", which includes a theory of gravitation. This involves much very abstruse and difficult mathematics, and there is much of it I do not profess to understand. I have, however, a profound admiration for the masterly way in which he has attacked a problem of transcendent difficulty.

APPENDIX

(i) List of my Cavendish Students who became Fellows of the Royal Society

The first of my pupils to receive the distinction of the F.R.S. was Callendar, who was elected to the Society in 1894 ; he was followed in 1897 by C. Chree, now deceased, who made important investigations in terrestrial magnetism for which he was awarded the Hughes Medal of the Royal Society ; in 1899 R. Threlfall (see p. 117) was elected ; in 1900 C. T. R. Wilson, about whose work much has been said already, and for which he received the Hughes, Royal and Copley Medals, and was awarded the Nobel Prize, was elected ; in 1901 W. C. Dampier Whetham (now Sir William Dampier) was elected for his work on electrolysis. In 1903 E. Rutherford (now Lord Rutherford and Cavendish Professor at Cambridge) was elected for his research on radio-activity which laid the foundations of that science, and for which he received afterwards the Rumford and Copley Medals and also a Nobel Prize. In the same year J. S. E. Townsend (now Wykeham Professor of Physics at Oxford and who came to the Laboratory on the same day as Rutherford), was elected for his researches on the ionisation of gases, which he has carried on with great assiduity and success ever since and for which he was awarded the Hughes Medal. In 1905 the Hon. R. J. Strutt (now Lord Rayleigh) was elected for his work on the discharge of electricity through gases and on the radio-activity of various rocks and minerals. He has also received the Rumford Medal. In the same year G. F. C. Searle was elected for his work on magnetic measurements and on electromagnetic theory. In 1906 H. A. Wilson was elected for researches which covered a wide range of subjects, especially those relating to the conductivity of flames. In 1909 one of the first research students, J. A. McClelland (now deceased), was elected for his research on the discharge of electricity through gases, and on Lenard rays; in 1912 C. G. Barkla (now Professor of Natural Philosophy in the University of Edinburgh) was elected for his discovery and investigation of the X-ray radiation characteristic of different bodies. He afterwards received the Hughes Medal for this work and was awarded a Nobel Prize. In 1913 O. W. Richardson, another old student who also later received a Nobel Prize and also the Royal and Hughes Medals, was elected for his researches on the emission of electricity from hot bodies, which are the foundation of the science of thermionics. In the same year W. Rosenhain (now deceased), who became Superintendent of the Department in Metallurgy in the National Physical Laboratory, was elected. In 1914 S. W. J. Smith (now Professor of Physics in the University of Birmingham); in 1915 J. C. McLennan (now deceased), late Professor of Physics in the University of Toronto; in 1917 J. W. Nicholson (formerly Professor of Mathematics in the University of London); in 1918 E. Gold, Assistant Director of the Meteorological Office; in 1919 G. I. Taylor, Yarrow Professor of the Royal Society; in 1921 A. A. Robb, the author of *Time and Space*, and F. W. Aston, who has received a Hughes Medal and also a Nobel Prize; in 1922 G. A. Schott (sometime Professor of Applied Mathematics in the University of Wales) ; in 1923 F. Horton (now Professor of Physics in Holloway College) ; in 1925 R. Whiddington (now Cavendish Professor in the University of Leeds) ; in 1926 L. F. Richardson, Principal of the Technical College, Paisley ; in 1927 E. V. Appleton, lately Professor of Physics, King's College, London, now Jacksonian Professor, Cambridge University, and recipient of a Royal Medal ; and in 1930 my son, G. P. Thomson, Professor of Physics in the Imperial College of Science and Technology, were elected Fellows of the Society.

(ii) *List of Universities and learned Societies in which my pupils have held Professorships*

(*N.B.*—When more than one Cavendish student has held the Professorship, the number is enclosed in brackets.)

England

> Cambridge (4), Oxford, London University—University College, King's College (5), Imperial College of Science (4), Royal Holloway College (4), University of Manchester (2), University of Liverpool, University of Birmingham, University of Leeds, University of Reading, University of Durham, Armstrong College, Newcastle, Royal Society, Yarrow Professors (2), Royal Institution (2).

Scotland

> Universities of Edinburgh, Glasgow (2), Aberdeen, St. Andrews.

Wales

> University of Bangor, Aberystwyth (2).

Ireland

Trinity College, Dublin, National University of Ireland (2), University of Galway.

Canada

University of Montreal (3), University of Toronto, Queens University, Kingston, University of Saskatchewan, Hamilton University of Ontario.

Newfoundland

Dalhousie.

South Africa

University of Johannesburg, Gray University, Bloemfontein.

India

University of Calcutta, Lucknow.

U.S.A.

Harvard, Princeton (5), Yale (2), Columbia, Minnesota, Texas, Cincinnati (2), Nebraski, Illinois.

France

Institut de Physique et de Chimie.

Austria

Vienna.

Germany

Frankfort.

Norway

Oslo.

Russia

Leningrad.

Poland

Cracow, Lemberg.

INDEX OF NAMES

INDEX OF NAMES

INDEX OF NAMES

INDEX OF NAMES

INDEX OF SUBJECTS

INDEX OF SUBJECTS

INDEX OF SUBJECTS

Printed in the United States
By Bookmasters